Praise for *Y: The Desc*

'Steve Jones' ideas drive me almost mad with wonder. He is a great elucidator, educator, teacher and thinker. Men, it turns out, are not from the planet Mars at all, but more magically and prosaically from Y, the differential chromosome that makes us geezers into the fantastic things we know we are and women adore us for it' Bob Geldof

'It's not so much Woody Allen's *Everything You Always Wanted to Know About Sex But Were Afraid to Ask* . . . more about a thousand and one things you never knew about sex and could never have imagined asking . . . required reading for anyone possessing the relevant chromosome – and XX partners, actual or potential, might find it rather helpful too' Steven Rose, *Guardian*

'An engaging tour of the male body and all its fluids, written with a style that is more than capable of competing for the attention of testosterone-addled minds . . . it enlightens and explains masculinity in a way that the therapists and theorists who have tried to define the subject could never do' Mark Henderson, *The Times*

'A sure-fire hit . . . Jones is a master storyteller who spices his books with wit and detail' Steve Connor, *Independent*

'The joy of this book is in the detail of male vicissitudes, from conception onwards. It is stacked full of wonderful anecdotes and vignettes' Jeffery Weeks, *Times Higher Education Supplement*

'A rattling good read . . . wide-ranging, amusing, deeply interesting and, so far as I can judge, authoritative' George Rosie, *Sunday Herald*

'Connoisseurs of the Jones' oeuvre will know what to expect – a lively, witty, sardonic tour of his chosen terrain' Graham Farmelo, *Lancet*

'A wealth of fascinating detail . . . and much delightful trivia' Fay Weldon, *Telegraph*

Steve Jones is Professor of Genetics at University College London, and has worked at universities in the USA, Australia and Africa. He gave the BBC Reith Lectures in 1991, and presented a BBC TV series on human genetics and evolution in 1996. He is a regular columnist for the *Daily Telegraph* and frequently appears on radio and television. His previous books include *The Language of the Genes* (winner of the 1994 Rhône-Poulenc science Book Prize), *In the Blood* (shortlisted for the 1997 Rhône-Poulenc), *Almost Like a Whale* (winner of the 1999 BP Natural World Book Prize), *Darwin's Island* and, most recently, *The Serpent's Promise*. In 1997 he won the Royal Society Faraday Medal for the Public Understanding of Science.

Y

The Descent of Men

Steve Jones

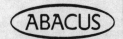

To Alex Trench who has at least one
and Norma Percy who doesn't

ABACUS

First published in Great Britain by Little, Brown in 2002
This edition published by Abacus in 2003
Reprinted 2005, 2008, 2009, 2011, 2013, 2014

Permission to quote from 'The Tower' from *The Tower* (1928)
and 'The Spur' from *New Poems* (1938) by W. B. Yeats has been
granted by A. P. Watt Ltd on behalf of Michael B. Yeats

A CIP catalogue record for this book
is available from the British Library.

ISBN 978-0-349-11389-0

Typeset in Bembo by
Palimpsest Book Production Limited, Polmont, Stirlingshire
Printed and bound in Great Britain by Clays Ltd, St Ives plc

Papers used by Abacus are from well-managed forests
and other responsible sources.

MIX
Paper from
responsible sources
FSC® C104740
www.fsc.org

Abacus
An imprint of
Little, Brown Book Group
100 Victoria Embankment
London EC4Y 0DY

An Hachette UK Company
www.hachette.co.uk

www.littlebrown.co.uk

Contents

'Ignorance more frequently begets confidence than does knowledge.'

Charles Darwin,
The Descent of Man and Selection in Relation to Sex (1871)

PREFACE

THE DIMINISHED FEMALE

Type the term 'masculinity' into Amazon.com (not at first sight the most obvious source) and a thousand titles emerge. A quarter are by women. They dissect the hapless male in volumes that range from the baffling – *The Penis Book: An Owner's Manual* by Margaret Gore – to the bilious (*If Men Could Talk, Here's What They'd Say*) and take in on the way examples from beastly (*Men and Other Reptiles*) to botanical (*Why Cucumbers Are Better than Men*).

Half a century after Simone de Beauvoir claimed in *The Second Sex* – a work, needless to say, about women – that 'A man would never set out to write a book on the peculiar situation of the human male', the first sex (if men can still so be described) has become a literary genre. Most of it reflects a potent sense of separation, fuelled by genes.

Men are, of course, different. They have small reproductive cells, for a start. Males copulate about fifty billion times a year, which comes to around a million litres of semen a day (that may sound a lot, but is a flow no greater than the fledgling Thames a few miles from its source). Every second, in return for their two hundred thousand billion global sperm, they are rewarded with five births. The world's women contribute a mere four hundred eggs with each tick of the clock for the same result. That cellular imbalance is at the centre of maleness. It confers on males a simpler sex life than their partners, together with a host of incidental idiosyncracies, from more suicide, cancer and billionaires to rather less hair on the top of the head.

Women are distinct in other ways as well, as much of fiction shows. Real devotees of the undigested argument from nature see society itself as a byproduct of the genital battle. Even those less keen to concede politics to evolution accept that many of the contrasts between the sexes, in brain and in body, reflect the divergent tactics of each contender as they struggle to pass on their genes.

The study of human evolution began in 1871 with Charles Darwin's *The Descent of Man*. He showed that our species – man, to give *Homo sapiens* a convenient label – had evolved from apes. His claim is now a truth universally accepted except by those determined to delude themselves. Men, in their descent as an entity distinct from women, have in comparison been neglected, but at the turn of the twenty-first century a new science of maleness has emerged, almost unnoticed. This volume sets out to explore it.

Its title, *Y*, laconic as it might be, refers to his special chromosome, the vessel of manhood. Half the population is a slave to its insistent presence. The cellular difference between males and females was discovered just a hundred years ago, in 1902, soon after Mendel's own work – the much-ignored experiments on peas which founded the scientific study of inheritance – had been brought back to light. In the brief century since then man's meagre genetic cargo has been unpacked and is now on view for all to see.

To use a single letter of the alphabet as the title for a book is a notion stolen from Pauline Réage's great work of sadistic pornography, *The Story of O*, a novel set in a chateau filled with women but run entirely for the benefit of men. My own exploration of their uneasy state emerges from a similar social milieu, but is, unlike Miss Réage's monosyllabic tale, the least erotic volume ever written about sex, for its raw material comes from science, with its ability to transform magic into prosaic truth, and not from simple erotic fantasy or the vast (and almost as fantastical) literature on the supposed dilemmas faced by men.

This volume marks my second attempt to act as Darwin's ghostwriter, for I once tried (perhaps it was a mistake) to update his greatest work, *The Origin of Species* and to follow what he called its 'long argument' step by step. Here I have a less ambitious aim.

The Descent of Man (with its extensive internal digression on *Selection in Relation to Sex*) was a series of diverse and marvellous ideas, many of them almost unsupported by evidence (the first Neanderthal had been found twenty years before but most people thought him to be the remains of a soldier who had died in Napoleon's retreat from Moscow). In the book, Darwin came up with the ideas that formed the science of *Homo sapiens*: our links with apes, the ties between animal and human behaviour, the ratio of the sexes, the evolution of lust, and the problem of body hair. My own work is not so daring. It is based on the single, simple and perhaps banal notion of the masculine. Although it does not follow the plot of *The Descent of Man* in detail, on the way it explores many of the questions raised by Darwin and calls upon a range of facts that would amaze and delight the author of its great original.

I assume some base of knowledge. As the seventeenth-century herbalist Nicholas Culpeper wrote in his *The Genitals of Men*, 'Latins have invented many names for the *yard* . . . I intend not to spend time in rehearsing the names, and as little about its form and situation, which are both well known, it being the least part of my intent to tell people what they know, but teach them what they know not.' The 'yard' in this context is the penis, and the vulgar details of its use remain unreported here. Darwin himself, faced with a similar difficulty, retreated into Latin, the language of educated men, but not of their wives and servants. To do so now would succeed too well in hiding the truth from untutored eyes. Instead I hope to use plain English to tell my readers at least something new about the past, present and future (gloomy as it may be) of those who bear the organ.

In Charles Darwin's time biology had scarcely begun. Sex was not discussed in public, and the upper classes at least remained uninformed until the last possible moment. Queen Victoria, who married in 1840, wrote to Lord Melbourne, her prime minister, describing her first encounter with her new husband as 'a most gratifying and bewildering night . . . We did not sleep very much'. Even a century later, at a time of great scientific progress, ignorance was widespread. An American biologist (whose first paper was entitled 'What Do Birds Do When It Rains?') showed how innocent his generation remained. In 1939 Alfred Kinsey, founder of the Institute for Sex Research at the University of Indiana, discovered that just one in five local teenagers knew that babies had a mother. None realised that a male was involved (and some supposed that an egg was provided by the hospital). I hope at least to put those misconceptions right.

Now biology has become part of popular culture. As a result, it faces another problem. For many people, the science of life explains, or so it seems, all we need to know about our sexual and even our social selves. Science is, without doubt, a crude, pragmatic but effective way in which to understand the world. The physicist Niels Bohr likened it to washing up after dinner: dirty crockery, a dirty cloth and dirty water but – as if by magic – clean dishes emerge (philosophy, he said, was the same but without the water). Unfortunately, the public has too much faith in what science can do. The arts faculty prefers the top-down approach – passion first, facts later – but for scientists sex has always worked better from the bottom up. Biology cannot say much about man's struggle to find himself or somebody to share a bed with, and most biologists do not bother to try. Even so, as it puts together the machinery, bit by bit, from genes, to bodies, to brains – and perhaps, some day, to emotions – science has at last begun to understand what it means not to be a woman.

To do so needs evidence from fields as diverse as genetics, anatomy and psychology. I have not been able to explore it all, but a biologist in the bedroom can take an objective look at males and soon finds much that is unfamiliar.

Once, female genitals were seen as no more than those of a male turned inside out. Such a view is naive, but the complete sequence of the human genome shows how tiny the differences are between husband and wife and how simple the mysteries of manhood appear when compared to the wider enigmas of life. Today's new insights put an end to the ancient belief in the masculine as the feminine form perfected. Humans have about fifty thousand genes, but just one in a thousand is unique to my fellow men. From their command centre on the Y chromosome they switch the embryo from its first and feminine state onto the rocky road to manhood.

From hormones to hydraulics and from homosexuality to hair-loss, man's mechanism has been deconstructed. Simple as it may seem, different creatures build it in quite different ways. Some use single genes and some chromosomes, while others prefer a sudden shift of identity whenever the chance arises. Males are in flux in almost every way: in how they look and how they behave, of course; but, more important, in how they are made. From the greenest of algae to the most blue-blooded of aristocrats their restless state hints at an endless race in which males pursue but females escape.

That contest, over billions of years, provides an infinity of images of the problems faced by makers of sperm. Taken one by one they are little more than anecdote, but in their entirety they reveal where males come from, how they cope with their uneasy state and even where they might be going.

Certain bedbugs insert sperm into their spouse through

her body wall. Their fellows improve on the process with a homosexual assault on the copulator, who passes the attacker's cells on to his next partner. Angler fish have a different balance of power, for each female is decorated with half a dozen sets of her partners' genitals. Their owners have wilted away, and she chooses which penis to use. Slime moulds, on the other hand, have thirteen sexes, twelve of them male.

Such tales are mere hints of how far the universe of manhood has been explored. Why do Viagra users see the world in blue? Why the April peak in condoms in the sewers of Belfast? Who castrated the Bishop of Lincoln? And why do men invariably claim more partners than women (which, given the laws of arithmetic, is impossible)? All these questions, and more, are answered by the new biology.

History, in its official version at least, also turns on great men. Pedigree hunters try to attach themselves to such lost figures and pursue their own forefathers in a chase that may become an obsession. The Y is the answer to their prayers. It identifies black descendants of the author of the Declaration of Independence and shows how the ancient warriors of the British islands were overwhelmed by a wave of alien females. It even ties today's Welshmen to the natives of Patagonia, the people who led Darwin, on his voyage on HMS *Beagle*, to the idea that humans descend from remote ancestors and, in the end, to write *The Descent of Man*.

As science uncovers man's past, its sister, technology, has had drastic effects on his future. For a husband suspicious of his wife's fidelity, the CheckMate kit will, for $49.95, test sheets with a dye that turns purple when semen is present. The Forensex company ('Send us your dirty panties!') does the job more discreetly. The impotent, like the cuckolded, can be helped, with pumps, rods and – for those with good eyesight and a certain indifference to pain – with injections into their wayward organ. In 1998 the pharmaceutical company Pfizer opened the floodgates to the world's penises (and, the *Sun* newspaper claimed, also stopped cut flowers from drooping; an assertion made in the spirit of vulgarity that later turned out to be true). Their blue pill has, half a billion prescriptions on, been followed by a new generation of erectile chemicals. When pharmacy fails, progress in artificial fertilisation is such that in some places one child in twenty is born after a liaison in a clinic rather than a double bed.

Sperm evolved to carry genes but is now used to move foreign DNA into eggs (the main success so far has been to make mice that glow in the dark). Human proteins are made in animal semen, which turns the litre of ejaculate produced by a boar into a useful commodity. Man himself may in the end become redundant, for his sperm can be grown in animal testes, and in mice at least an egg can be fertilised with a body cell from another female, which cuts out the secondary sex altogether.

Technology, in its triumphs, began before science and has done great things without its help. Sociology is younger, and less confident about its foundations. Once, it turned to psychology or politics as an alibi. Now it has become obsessed with biology as a universal explanation of human affairs. Perhaps, some say, Darwin will succeed where Freud ('anatomy is destiny') and Marx ('the real point is to do away with the status of women as mere instruments of production!') have failed. So far, the hope has been in vain. Our paternity may descend from ancestors shared with chimps and gorillas but our habits as social but rather monogamous animals are closer to those of seagulls than of any primate. Terms such as 'gender identity', 'sex stereotype', 'transsexual', 'transgender', 'masculine', 'feminine' and the rest accept the confusion of human lives compared to the simplicity of those of animals. Science is happy to try to understand the two sexes, and to leave it to society to recognise as many genders, identities and roles as it likes.

Maleness is simple but manhood is infinitely complex. The condition is much discussed in the United States, where the politics of inequality that occupies the rest of the world has to a great extent been forgotten. The American Men's Studies Association lists fifty university courses in their subject: an impressive total, but a tenth the number devoted to the other half of the populace. As Gibbon noted in *The Decline and Fall of the Roman Empire*, 'In every age and country, the wiser, or at least the stronger, of the two sexes, has usurped the powers of the state, and confined the other to the cares and pleasures of domestic life.' Maleness is no more an excuse for injustice than is wealth but I leave its politics aside.

Theology, too, is obsessed with that section of society. Roger Goodland's great work *A Bibliography of Sex Rites and Customs* has a thousand references to their doctrinal importance ('Circumcision bonnets; Penis, Ablation of, as sacrifice; Priests, etc., Coitus with, thought sacred'; and many more). This literature, too, I have decided – with some regret – to avoid.

Much as scholars of politics, of society and of religion ignore the limitations of science, biologists themselves can be ambiguous about where their subject ends. A third of today's research papers on topics such as bird behaviour or testosterone use 'gender' rather than 'sex' in their titles. That is absurd, for the message of science is clear. Animals have males, but only *Homo sapiens* has manhood. As a result, genes say a great deal about sex, but rather little about gender. The failure to separate the two confuses those who feel that biology can explain everything (or nothing) about those blessed or otherwise with a Y chromosome.

Charles Darwin, in the last paragraphs of *The Descent of Man*, was less than charitable about his experiences in South America on the *Beagle* voyage, forty years before: 'The astonishment which I felt on first seeing a party of Fuegians on a wild and broken shore will never be forgotten by me, for the reflection at once rushed into my mind – such were our ancestors. These men were absolutely naked and bedaubed with paint, their long hair was tangled, their mouths frothed with excitement, and their expression was wild, startled and distrustful . . . He who has seen a savage in his native land will not feel much shame, if forced to acknowledge that the blood of some more humble creature flows in his veins.'

Now the notion that we share our blood with creatures much older, if not humbler, than ourselves, is universal, and lies at the basis of the study of our own species. However, many of the dilemmas faced by my own segment of the population are not a matter of science but of image. Darwin himself fell foul of the Victorian assumption that different meant worse and that savages had risen but little from apes. Today's sexual stereotypes are, in the same way, so familiar that their lack of substance is easy to forget. As someone who fails so singularly to live up to them, I hope to redress the balance.

A male geneticist may have some of the skills needed to write about the genetics of men, but in many senses I do not make the grade. I have never punched anyone (except when I was robbed some years ago on my front doorstep, when I broke several fingers at the expense of minor damage to my assailant's nose). I cannot ride a bicycle, remain childless, learned to swim at the age of forty and have never been to a football match (indeed, in five years of compulsory school sport not once, as far as I recall, did I touch the ball). I am as open to fantasies as the next man, but my dreams do not involve rape or torture – and I cannot even admit, with Woody Allen, to a certain fascination with chickens.

I have a lot to live up to. Steve Jones was the lead guitarist of the great punk group The Sex Pistols. His most popular Internet eponym is a bronzed Californian body builder with a gold posing-pouch, followed by a choice of Steve Jones as champion golfer, runner, yacht-designer, aerobat, footballer, boxer and Welsh national rugby player (with a famous hairdresser and a hat-maker to leaven the mix). Faced with such competition I stick to what I know: the biology and evolution of males.

Many people have helped me to write this book (and some even have a Y chromosome). They include Douda Bensasson, Tracy Chapman, Jim Cummins, Laura van Dam, David Goldstein, Greg Hurst, Bob Lieberman, Michael Morgan, Chris Pomery, Jenny Sadler, Saima Shah, Peter Tallack, Kay Taylor and Mark Thomas; together with Dave Elford of the UK Environment Agency who put right my rash assumption that mankind's output of semen was equivalent to the flow of the Thames at London Bridge. I thank them all here, and apologise – in as manly a way as possible – for failing, too often, to take their advice.

This volume should be of interest to at least half the population. It tells the story of the search for the nature of manhood and of the grudging recognition that the ancient view of females as diminished males has been reversed, to reveal instead each man's struggle not to be a woman. Those who prefer something a little more spicy have plenty of choice. *Superpotency: How to Get it, Use it, and Maintain it for a Lifetime* is out of print, but for advice from an expert I recommend Steve Jones's seminal work, *For Men Only: Winning at the Dating Game* (Trafford Publishing, 1997), still available from Amazon at $25.00.

CHAPTER 1

NATURE'S SOLE MISTAKE

Ejaculate, if you are so minded and equipped, into a glass of chilled Perrier. There you will see a formless object, but look hard enough – or at least so eighteenth-century biologists believed – and a baby appears; the male's gift to the female, whose only job is to incubate the child produced with so much labour by her mate. So central seemed a husband's role that his wife was a mere seedbed; a step below him in society, in the household and, most of all, in herself.

Foolish of course and quite wrong, for biology proves that man, and not woman, is the second sex. His sole task is to fecundate his spouse, but quite why he does it remains a mystery. To divide is more efficient than to unite and everyone has a history of a single sexual event when sperm met egg, followed by billions of cell divisions without its benefit. Untold numbers of species manage without even that masculine moment and for most of the time do not seem to mourn its absence.

Why men and why so many? Surely a few or even one would do, yet males are everywhere and do not always behave well. As Lady Psyche in Gilbert and Sullivan's operetta *Princess Ida* sings, 'Man is coarse and Man is plain/Man is more or less insane/Man's a ribald/Man's a rake/Man is Nature's sole mistake!' Much of modern biology is an affirmation of her claims.

Man's state can be defined in several ways, but most are frivolous. Those who claim it are indeed plain, and coarse, and possessed of a penis. None of those qualities is very significant, given the bizarre appearance of half the members of many species and the many ways invented by evolution to deliver their sex cells. Men themselves have a special structure that shunts them towards their fate. As well as the twenty-two chromosomes shared by each sex, women have two large X chromosomes, while their partners have a single X chromosome paired with a smaller Y. Fundamental as it might seem, the Y is not the root of maleness, as other creatures gain the state without chromosomes at all. Even ribaldry and rakishness are not peculiar to one partner, for plenty of animals leave the husband to hold the baby while a wife searches for a new mate.

To biologists, masculinity turns only on the size of the sex cells. Such things come in large and small varieties and the males make the small ones. They put their bets on an outsider: on a single winner among billions at the post, each stripped down ready to face a risky gallop to the line. Their spouses, in contrast, stake their all on a few more-or-less safe bets. Every egg has a fair chance of a plod around the sexual racecourse, but each carries, as a massive weight penalty, the goods needed to make an embryo. Those who make sperm take a free ride at the expense of their opposite numbers, for men – by definition – do not give birth. Instead, they use female flesh to copy their own DNA.

Males act in their own interests but as an incidental perform a vital role in evolution, for they act as conduits through which genes move between females. Without their help, all new mutations would be confined to the direct descendants of the individual in which they arise and life would at once become a multitude of clones rather than a set of unstable biological alliances formed anew each time sperm meets egg.

Men bring women together. They make links between families and allow genes to be tested against nature in new and perhaps fruitful coalitions. Expensive as they are, once evolved such creatures are almost impossible to get rid of. A certain group of tiny freshwater animals managed to do away with them a hundred million years ago, but for all others a burst of maleness is needed now and again.

Its humble task flies in the face of tradition, which has long seen man as the bearer of the human heritage and his mate as, at best, a tiresome detail. At Karnak, in Upper Egypt, the great god Amun gave life to the world by masturbating over it. In a more decorous myth, Genesis allows Eve a mention, but by chapter five she too has gone: 'And Adam had lived an hundred and thirty years, and begat a son in his own likeness, after his image; and called his name Seth.' After her moment of glory in the first book of the Bible (one of the few written before the emergence of an all-male priesthood) her sex fades away.

Eve's consort could not, of course, reproduce without help as he was, ribs apart, a blind alley. What makes Adam and his descendants what they are?

Physics was transformed in 1905 with the theory of relativity. Everyone knows who thought of that idea, but Nettie Maria Stevens, the Albert Einstein of manhood, is forgotten. Like Einstein, she started outside science and turned late to research. In the year of relativity, when sex chromosomes seemed no more central to masculinity than is a moustache, Miss Stevens explained how they work. Flour beetle sperm came, she found, in two types, one with a large version of a certain chromosome, and the other with a small, the famous Y. The truth about maleness was revealed.

Two centuries earlier, the Secretary of the Royal Society had recorded a remarkable case: 'A country Labourer, living not far from *Euston-Hall* in Suffolk shewed a Boy (his Son) about fourteen Years of Age, having a cuticular Distemper. His mother had received no fright . . . his Skin was clear at birth – by degrees it turned black, and in a little time afterwards thickened, and grew into that State it appeared at present.' The boy, Edward Lambert, exhibited himself in London: 'To be seen at the George in Fenchurch Street a Man and his Son, that are cover'd from Head to Foot with solid Quills, except their Face, the Palms of their Hands, and Bottoms of their Feet.' He had many children and grandchildren. They, too, were exhibitionists, although some found it necessary to decorate their story: 'The young man is . . . *covered with Scales* . . . nearly half an inch long, and so hard and firm that with the touch of a finger they make a sound like stones striking together . . . the great-grandfather of the singular family to which this young man belongs, was found savage in the woods of North America.'

The Lamberts are much quoted as an example of a condition that passes from fathers to sons. Even Darwin mentions them. Sadly, the records reveal scaly daughters, too, as proof that their state is not in fact coded for on the Y chromosome.

Several genes have professed that noble home but not for almost a century after its discovery did one earn it. Some of the false claims arise because, in small families, a trait may appear by chance in sons alone. Others are due to mere bias, as a history of maleness has, for men, a great allure. In a certain Indian lineage, all the men – but none of the women – have points of hair (sometimes waxed by their proud bearers) on their ears. I once met a hairy-eared African who, when I asked whether any of his relatives shared it, told me his mother did. Most other supposed cases of Y-based inheritance also collapse on closer inspection.

Now everything has changed. The Y has come into its own. Biology no longer needs freaks of nature to track down genes. Instead it can go straight to the DNA. The completion of the entire human sequence, with its great string of chemical bases of four distinct types, has transformed the image of men. Our biological heritage, the Human Genome Project showed, is filled with parasites, redundancy and decay.

The chromosome unique to men is a microscopic metaphor of those who bear it, for it is the most decayed, redundant and parasitic of the lot.

The male badge of identity is small indeed – just a fiftieth of the total genome, with sixty million or so of the three thousand three hundred million base pairs in the entire sequence. Three-quarters of the double helix as a whole consists of spaces between genes, and the genes themselves contain hundreds of redundant sections. Much of the DNA exists as duplicates, multiplied again and again, with the copies diverged into families. It is also marked by innumerable segments of foreign material that have elbowed their way in. They are matched by internal hangers-on and by other great portions that have gone to rack and ruin. As a result, just a few parts in a hundred bear useful information.

Masculine decadence is such that, on the Y, a mere one piece in thousands does so. Every gene has a molecular signature that sets it apart from the material around it. The Y has but seventy or so segments that code for proteins, compared to ten times as many on the X. More than half consists of multiples of two uninvited guests and most of its duplicated elements are no more than corpses. In spite of some local hints of order, man's defining structure is a haven for degenerates.

It has a single redeeming feature. To half the human race the Y is the prince of chromosomes, for it gives the embryo a testis. There resides the noblest of all genes, the *sine qua non* of maleness. The crucial section sits near the structure's tip and, in its absence, a fertilised egg becomes a female. It is – at first sight – simple.

The key to man's nature came from some unusual men. Armed with penises though they are, such people lack a discernible male chromosome and carry (like women) two copies of the X. Their predicament comes from a genetic accident. A tiny portion of their father's Y was broken off when his sperm was made, and became attached to his own

X chromosome. Eggs fertilised by X-bearing sperm are expected to develop into girls, but in this case the uninvited passenger brought as a guest an extra length of DNA. It includes the gallant structure that impels a baby into boyhood.

The *SRY* gene itself (the initials stand for sex-determining region of the Y) was tracked down in 1990 after a hunt through that nomadic fragment. It contains fewer than a thousand DNA bases, which code for a mere 204 amino acids (the units bolted together end-to-end to build proteins). Unlike most genes it has no inserted sequences of useless material. *SRY* is small but potent. When a copy was injected into a normal XX mouse egg, the young animal that emerged (Randy, by name) was anything but female. His sisters gladly accepted him as a mate, to give a consumer's seal of approval for the power of that tiny gene.

Modern genetics takes place not just in animals (*in vivo*, as biologists say), or even *in vitro*, in test tubes, but *in silico*, inside computers able to match DNA sequences with each other. A search for genes with some resemblance to the master of manhood reveals a host of relatives, with the most similar of all upon the X, where it is active in the young brain (which might have interested Freud).

The monarch of maleness belongs, the computers show, to a large and diverse family. Its members do many things (and even control the reproductive lives of mushrooms). Each is in charge of some aspect of development and all share a special sequence of eighty or so amino acids. This forms a groove in the molecule which allows it to bind to a variety of DNA sequences. As it does, the double helix bends and, somehow, its change in shape turns on the target gene. At once, the machinery of growth springs into life.

SRY is a switch that directs other genes onto their allotted path. Like the railway points outside a large terminus, the testis-determining element guides, with a single tiny shift, the sexual express train towards one destination rather than another. It is (with a few rare exceptions) the sole gene absolutely necessary for a testis to be made, although to make the organ itself and the other useful structures that decorate all males needs hundreds of others, scattered all over the genome. *SRY* kicks into action four weeks or so after the egg is fertilised. In its brief moment of glory it sends billions of babies on a masculine journey. Quite how it does so, nobody knows, as its prime target – the leading wheel, as it were, of the embryonic locomotive – has not yet been found.

Most of its fellow passengers are associated with maleness. In guppies the gaudiest animals attract the most mates – and the genes for bright colour are on the male chromosome. Our own version is involved in the manufacture of sperm, in the rate of growth, in the formation of teeth and of certain brain proteins, in left-handedness and – if mice are a guide – in aggression. Its few other useful sections do the day-to-day jobs needed by all cells.

Outside the segments devoted to these small tasks, most of the Y is filled with decay. It has degenerated because it abjures the messy business of sex.

To biologists, that pastime is simple. It adds statistics to nature, for without it every child would be an exact copy of its parent. Copulation causes random noise in the world of the double helix, because it mixes up genes. Without it evolution could hardly happen. Chromosomes are present in double copy in most cells, but sperm and egg each contain just half the DNA of the person who made them. As a result, just one member of each chromosome pair can get in (which is why half of all sperm – made as they are by XY individuals – have an X and half a Y). By chance, a child may receive its mother's edition of some chromosomes and its father's of others. The process (recombination as it is called) goes further, as it reorders their very material. The members of each pair line up as sperm and eggs are made and then intertwine, break and rejoin in novel ways.

From a gene's point of view, reshuffling of this kind is a great restorative, as it allows it to escape from its neighbours – and to move house can be a great help when the individual next door is a feeble character who might, when exposed to the rigours of the world, drag a whole block down to his own level. Recombination means that sex, not death, is the great leveller. It allows new and hopeful blends to appear each generation and can get rid of several damaged pieces of DNA at once if they get into the same sperm or egg. The recipient becomes a scapegoat, for his demise purges several inborn sins at the same time. In his lonely fate he rescues many of his fellows who might otherwise be condemned by inferior genes.

The Y, in its solitary state, disapproves of such laxity. Apart from small parts near each tip which line up with a shared section of the X, it stands aloof from the great DNA swap. Its genes, such as they are, remain in purdah as the generations succeed. As a result, each Y is a genetic republic, insulated from the outside world. Like most closed societies it becomes both selfish and wasteful. Every lineage evolves an identity of its own which, quite often, collapses under the weight of its own inborn weaknesses.

Celibacy has ruined man's chromosome. The Y is a dead end for DNA as, once on board, it is impossible to escape through the hatch marked 'sex'. Most genes can use recombination as an emergency exit to flee, with a set of new companions, into the biological lifeboats known as sperm and egg. Only the Y lacks such a release. As a result, it has become confined to a single narrow interest: maleness. Any gene able to help in the task is welcomed but all others (apart from those involved in its own internal economy) rot away. Males evolved to stop females from degenerating into clones, but in their own intimate selves have suffered the same fate.

Where did the Y come from, and why is it so different from its fellows?

It bears the scars of a complex and unexpected past. DNA hints that the crucial structure was once the equal of its larger associate, the X, which now contains far more genes. It has suffered many ups and downs since the break with its partner. Both halves of this odd couple have altered over time, but one version has changed at breakneck speed. It has lost thousands of genes and gained just a few. As a result, the Y is an arriviste on the evolutionary scene with a unique identity of its own.

Fossils prove when, in the distant past, various groups of mammals split apart. The sex chromosomes of a series of our more and more distant relatives, from chimps to mice to wallabies, also show how they have changed through time. Their pattern of evolution is a hint of the active, bitter and ancient battle of the sexes that has driven males since they began.

When it comes to courtship or childcare, the interests of each parent differ in obvious ways. The chequered history of a chromosome shows that the tension between them goes far deeper.

Males are, in many ways, parasites upon their partners. Their interests are to persuade the other party to invest in reproduction, while doing as little as they can themselves. Like all vermin, from viruses to tapeworms, they force their reluctant landladies to adapt or to be overwhelmed. As the host evolves to cope with her unwelcome visitor, the two parties enter a biological dance. Each has its own agenda and, as one gains, the other fights back. Quite often, the evolutionary pas de deux takes up a frenzied pace.

New parasites evolve all the time, and the older kinds often change their identity. Most of our diseases are recent and can be traced to an enemy that began in an animal (as is the case for the agent of AIDS, which came from apes, and of malaria, from birds). Whenever a new illness emerges, or an old ailment flares up, those under attack must respond or die. The parasite, whatever it may be, constantly tests its host's fortifications with new mechanisms of virulence – which in turn are fended off. Neither party can afford to relax and, little by little, a complex and rickety structure of defence and counterattack evolves.

Like the front line in the First World War, the evidence of past assaults may make no apparent sense and can even sink from view. The foes may stay immobile for years until a sudden advance by one forces a response by the other. The battle of the parasites – and the sexes – anticipated the strategy of Generals Haig and Ludendorff by millions of years.

Every year I take a group of undergraduates on a field course to Spain to study evolution in action. On the streams, water-skaters swim in pairs, with the male on top. His embrace comes not from affection (as the old texts have it) but is a desperate attempt to keep off his rivals, with thousands of attacks before a second mate has any chance. As the struggle goes on, the object of his attentions begins to starve, but she cannot shake off her partner. In the meadows above, flowers attract insects to pollen – a male attribute – but the female parts of the plant abort almost all the seeds that result from their efforts. The students take notes and go to the bar.

The strategies of water-skaters and plants reflect the divergent interests of two factions. Monogamous creatures (and they are rare, and have become rarer as paternity tests reveal the sordid truth about nature) live in harmony, as any harm done to the prospects of one does the same damage to its partner. For all others, the interests of each player diverge. Males hope for another encounter, while their spouses must decide whether to nourish the results of an erotic fling or to wait for a better chance. For one participant, success turns on how many individuals can be inseminated, while the other is best served by a choice of the finest mate. The antagonism sets off a conflict as bitter as that between the British and German armies and leads to the spectacular rows among sea elephants or peacocks beloved of film-makers.

Blubber, plumes and the rest are mere vulgar brawls in a larger engagement. The image of males – once seen as the prime movers in the sexual universe – has faded in the face of the truth: that, for much of the time, their backs are to the wall. The dispute with those who copy their DNA forces their very being into flux.

Pasiphae, wife of Minos (who later gave birth to the Minotaur, a beast much painted by the spermatic enthusiast Pablo Picasso), was fed up with her husband's infidelity. She put paid to his mistress in an ingenious way. With a little necromancy she made him 'pour forth in his semen a swarm of venomous snakes, scorpions and centipedes, which devoured the woman's intestines'.

The Minotaur's mother was ahead of her time. Biology reveals how males can attack the molecular entrails of their mates and how their targets, as they try to escape, oblige them to update their armaments. Semen as both weapon and gift is a microcosm of the endless two-horse race that drives men and their mechanism forward.

Seminal fluid does many things. It carries sperm, of course, but can also compel those who receive it to invest more than they might like in those cells' success, whatever harm is done to their own prospects.

Insect sperm comes in a poisoned chalice. Some of the ejaculate's chemicals resemble digestive enzymes. They disable foreign sperm and increase the penetrative power of those of the male who makes them. Others are more subtle, for they act on females themselves. They plug her reproductive tract, force her to store the donation, reduce her desire to mate again, and oblige her to make more eggs than she might prefer. Some insects are selfish enough to impose monogamy on their partners while staying promiscuous themselves. A female housefly, for example, is limited to a brief encounter, as her partner's poisons are so potent that after a single sexual experience she can never mate again. Fruit flies are less efficient bullies, but can still damage their mates.

The evolution of the hundred or so specialised proteins found in insect semen reveals a series of repeated engagements in which attack is countered by defence. Some males are ten times better at displacing foreign sperm than are others. In the same way, animals with certain variants in the ejaculate can overcome the resistance mechanisms of particular mates, while other females can resist those individuals but yield to the assaults of another version of the seminal weapon.

The balance of terror is fine indeed. When flies are allowed to mate just once in a lifetime over many generations, seminal fluid evolves to become more benign, as, without a chance at a second mate, it pays its makers not to harm those who receive it. Females in turn become less defensive about the crucial but risky liquid. Once removed from this Garden of Eden and returned to the real world, neither party can compete: the tamed males are less able to fertilise a partner, and their newly confident mates may be, like the victim of Pasiphae, killed by foreign sperm.

The divergent interests of those who manufacture and those who accept sperm make seminal proteins change far faster than molecules with a less salacious task. Closely related species of fruit fly differ more in their ejaculate molecules than in any other. Even humans and chimps, similar in most of their DNA, are distinct in the sections responsible for such proteins, a strong hint of a history of conflict between body fluids in our own recent past.

Bitter as it may be, the battle of the secretions is little more than a skirmish in the war that moulds the essence of every male.

Males look different. All cock birds go in for courtship, but they do it in their individual ways. Canaries sing, peacocks have great tails, ruffs dance in natural ballrooms in front of their mates, and bowerbirds build shelters to impress their opposite numbers. Even close relatives can be quite unalike.

Female ducks all look rather similar, but their partners each have their own gaudy (and in the hunting season dangerous) displays. Their desires are unchanged, but each species advertises them in a unique and what seems an almost arbitrary fashion.

The control of sex is much the same. On the larger scale of evolution, males are made with distinct, and at first sight unrelated, strategies. Some creatures depend on genes and some on whole sets of chromosomes, while others turn to environmental cues to decide on which sexual identity to assume. In the grand reproductive handicap, stalemate is interrupted by change and, now and again, by a large and strategic shift that pushes matters into a new phase. Masculinity emerges as a fragile and uncertain thing which is often forced to re-invent itself. Whatever is hijacked to its ends, from tails to genes, begins at once to crumble away.

Parasites reduce themselves to a bare minimum as soon as they can. Faced with a host who shifts the goalposts each generation, they have to run to stay in the same place. It pays to be as unhindered as possible, which often means reduction to a mere sack of genitals and guts. They may shrink at the molecular level, too. The leprosy bacillus, for example, has lost hundreds of genes in comparison to its free-living ancestors. Man's most basic attribute also has a strong tendency to wilt.

Australian males are a model of virility, but the average native has a rather small version of what makes him what he is. Kangaroos split from our own ancestors a hundred and thirty million years ago, while the platypus made the breach even earlier. Both animals have tiny Y chromosomes, barely visible even under the microscope. Each has a mere ten thousand DNA bases (compared to sixty million on the human version) but, minute as it is, the entire antipodean Y can be paired up with a small part of our own. It contains the local version of the SRY gene, a core of masculinity left from the days when mammals began. Outside this tiny sector the male trademark of the kangaroo has been whittled away.

In parts of the northern hemisphere, the rot has gone further, for the mole vole attains masculinity without a Y chromosome or an *SRY* gene at all (how it does so, nobody knows). In a matching perversion of the rules, up to half the females of certain South American mice are XY − and they have more offspring than those who are satisfied to stick with two Xs. Although few mammals have gone quite so far towards downgrading the Y chromosome, most have suffered repeated changes of molecular personality, with much divergence even between close relatives. The viruses on our own version − unlike those elsewhere in the genome, which are shared with mice or flies − are different even from those of chimpanzees. Given the rate of decay since it began, the Y might disappear altogether within a mere ten million years.

The *SRY* gene itself has undergone many changes of personality. It does the same job in platypuses and primates, but just a small part is conserved. The central piece − the length that binds to DNA − is the same in kangaroos and humans, but elsewhere evolution has been busy. Among the mammals, *SRY* evolves ten times faster than the other members of its gene family. The version found in mice, for example, is twice the size of its equivalent in *Homo sapiens*.

About two thousand proteins have been read from end to end in both mice and men. When compared by computer, most differ by less than one in ten in the order of their building blocks. A hundred or so of the molecules stand apart, for they have diverged by half or more from their fellows. A substantial proportion of that group is involved in reproduction, while the rest is responsible for parts of the immune system and for the antiseptic protein found in teardrops, both of which are needed to deal with parasites.

The grand falling-out of the sex chromosomes began long before the evolution of men or mice − or mammals − themselves, soon after the separation of our own ancestors from those of birds, three hundred million years before the present. The Y traces its origin to a time long before the death of the dinosaurs, when tree ferns and simple reptiles ruled the land.

The DNA of one tip of the modern X resembles that of its partner, while the other end is quite distinct. Like a zipper as it opens, divergence has spread from one extremity of the two once-equal structures. The rupture happened in fits and starts, and the male chromosomes of a range of mammals have a stepwise, rather than a gradual, pattern of separation from their opposite numbers. In addition, vast blocks of genes have reversed their order. The first reshuffle took place when the ancestors of birds and mammals separated, whereas the last did not happen until the appearance of the hominid line a few tens of millions of years ago. Such upheavals have moved the *SRY* gene to the far end of the chromosome from its relative on the X.

The Y chromosome has had many other adventures on its journey into decline. Its DNA comes in two flavours. Certain sections have matches on their opposite number, but others find their kin elsewhere. The Y is a men's club for genes and some of its members come from afar.

Many men cannot make viable sperm (azoospermia, as the condition is known). The Y, like its owners, is fragile, and about one man in a hundred is infertile because of a new mutation, which is often manifest as a small break in the chromosome. The genes involved give an insight into the nature of maleness.

Women, with two identical sex chromosomes, might seem more symmetrical than their XY partners but in fact men are the more reflective party. The Y is, in parts, a great hall of molecular mirrors. Upon it live several genetical palindromes: immense and much-reversed lengths of DNA whose repeated and symmetrical sectors retain almost perfect matches across vast numbers of bases. 'Madam, I'm Adam' is a gentlemanly enough phrase, but its eleven letters are dwarfed by their molecular equivalents, which may contain three million DNA units. Nobody knows what preserves these long segments from decay but, somehow, males can keep them under control. The main genes involved in spermlessness sit within that region and almost all men with the problem suffer a break in just the same place.

A computer search reveals that some of the Y's twenty or so genes have equivalents on their ancestor, the X. Others do not. A few look like genes on other chromosomes, which is no surprise, for such structures often break and rejoin as evolution goes on. The migrants have suffered some sea changes since their arrival in the ocean of palindromes, but retain much of their identity. Other passengers, though, have reached their refuge in quite a different way.

To make any protein, a long messenger molecule is first read off from the DNA. Within it are both useful sections and various bits that seem to contain no information. These redundant parts are cut out by a set of special enzymes and the edited instructions (now without inserts) pass to the cellular factories. Why life is so profligate, nobody knows.

One group of Y-chromosome genes lacks those superfluous pieces. Its members exist as abridged versions of a set of relatives found elsewhere in the genome. Their reduced state hints at an unexpected history.

Mammalian DNA is besieged by viruses. Some hijack the revised messenger molecules on their way to the cellular factories. They insinuate themselves and their passenger into a new site. Several Y-based genes were delivered in this manner. The gene whose failure causes many cases of infertility is a recent arrival, for in lemurs, our distant primate relative, it is on another chromosome, while in monkeys, apes and ourselves the structure has made its way to the master of maleness. The great leap took place after the split between the ancestors of lemurs and humans, fifty million years ago, which is, to an evolutionist, the day before yesterday.

The Y, it seems, is a pragmatist, happy to welcome any immigrant – however it arrives – if it is useful but ready to ignore those who are not. Most of the settlers are unlucky. They make a protein of little value to the male and, without a job to do, decay.

Such endless strife has pushed the masculine state into flux on a scale far greater than the Y chromosome. It is achieved in quite dissimilar – and at first sight unrelated – ways in different creatures. Such shifts hint at great strategic advances and retreats in the struggle for masculinity.

In birds the world turns upside down. Males are, in effect, XX (with two large sex-determining elements) and their opposite numbers XY (with one large and one small). The bird 'X' and 'Y' do not match their human equivalents but are kin to a different chromosome (an ancestor of our own number nine), as proof of an independent origin. Snakes have a bird-like system, with XX males, but for them yet another chromosome pair has taken up the banner of virility.

Fruit flies achieve the manly state in a different way. Males are XY but a fly's identity depends not on a special gene but on a balance between the number of Xs and all others. Children born with just a single X (and no Y) are girls because they have no *SRY* to put them on the alternative track, but flies in such a predicament are male, because they have half the number of Xs as normal. Bees and wasps go even further in their definition of males as diminished females, for sons have half as many of all their chromosomes as do daughters.

Some plants have Y chromosomes, which must have arisen independently from our own and, as a final twist to the tale, certain creatures use schemes which at first sight fly in the face of our own familiar rules. For them, identity comes not from within, but from the world outside.

In the Middle Ages, life seemed simple: heat on the right side of the uterus gave a boy, on the left a girl. That eccentric notion is not altogether false, as in many fish and reptiles sex depends on the temperature at which an egg is incubated. The details, as so often, are filled with ambiguity. In alligators, heat makes sons, and cold makes daughters. In most turtles the opposite is true. Hot eggs grow quicker and make larger young. Alligator males gather a harem and it pays any male to be as big as possible – which is why they have sons in the warm. Many turtles (with sons in cooler nests) mate peaceably in the ocean, and large males do not gain much – but large females lay more eggs. To clinch the matter, snapping turtles (freshwater beasts who quarrel over females) return to the alligator pattern, with sons from warm eggs.

Sex responds to many other pressures. In a certain marine snail of banal tastes, a young animal lucky enough to meet a female becomes male, and vice versa. In many insects, high density leads to sons, and fish have a whole gamut of gender shifts which depend on the emotional rather than the physical temperature, as a female changes sex when left alone with a group of her fellows anxious to find a mate.

Although the end is reached with such a variety of means, maleness itself seems simple enough. Alligators, flies, peacocks, snails and fish are, in their shared passion for sperm, not much different from men themselves. Why is their state defined in such disparate ways? Sex and males have been around for much longer than people or alligators and it seems odd that their controls have changed so much.

The genes in charge of the basic body plan – left or right, up or down, front or back – are shared not just by men and women but by alligators, flies and worms. Different as such creatures might appear, all have the same basic layout. Once, its foundations were hidden, overlaid by the diversity of shape and size conjured up by evolution. Not until biologists dug deep into DNA did those very dissimilar animals reveal their hidden likeness.

Genetics has started to uncover the groundwork buried in the debris of sex. Because maleness changes at such speed, the rubble is piled higher over those ancient genes than in most places, but today's excavating equipment has begun to dig them out. They hint at what it really takes to be a man.

Hermes – messenger of the gods, controller of dreams and guardian of livestock – fell for Aphrodite, the goddess of love (who herself, as it happens, emerged from *aphrodes*, a sea of foaming semen flooding from her castrated father). Her first son, Priapus, was foul in appearance but her second – Hermaphrodite – was a boy of matchless beauty. As he bathed one day in a fountain, he was seen by its nymph. She begged to be allowed to unite with him – and her wish was granted. Their metaphorical descendants hint at the deep secrets of the priapic state.

One child in several thousand has both testes and ovaries (or, sometimes, tissues that mix the two). Some possess a Y chromosome, but most do not. Like normal women, they have a pair of Xs and, as a result, have no *SRY* gene. How can a fetus grow a testis when the crucial sequence of DNA is not there? *SRY* is not, it seems, quite such a *sine qua non* as it seems.

Hermaphrodites are asymmetrical in their ambiguity. Something has pushed their manhood to one side, for most of them have a testis on the right matched with an ovary on the left, and just a few show the opposite pattern. Such reproductive lopsidedness has to do with the rate of growth in the first days of development.

The right half of an embryo grows faster than its mirror image. What once seemed no more than a curiosity was in fact a hint at a profound truth: that genes far older than *SRY* once ruled the world of maleness. The fetal testis is, from its earliest days – well before any Y-based gene kicks into action – twice as big as its female equivalent. The gonad is given an initial push towards its fate by an ancient and hidden mechanism, to which hermaphrodites in their asymmetry hold the key. The male sex-determining gene is, like the antlers of deer or the tails of peacocks, more a symptom of masculinity than its ultimate cause. *SRY*, like the mechanisms with which birds, bees, alligators and the rest decide their state, stands on the shoulders of some earlier masters.

Sex, like politics, depends on a hierarchy of command. Empires collapse and are superseded, and masculinity is much the same. Ancient controls have been overlaid by new devices, quite different in different creatures – but, now and again, the old rulers issue a reminder of their existence. They are the ghosts of a world of maleness which evolved long before the animals who indulge in that pastime today.

Men, flies and worms have all had their entire DNA sequences deciphered. Distinct as they appear, all three possess rare and shared mutations in certain groups of genes that lie close to the roots of manhood. When damaged, they cause males to develop as members of the opposite sex – and they can exert their transforming power even when moved by clever biologists from flies to worms. The existence of a shared set of errors, with an ability to work in such different creatures, hints at an ancient foundation of masculinity which, should it be damaged, causes the whole structure that rests upon it to fail. Then, the other party hidden within can make an appearance.

A few XY babies develop as girls. Most have a normal *SRY* gene, and have – like flies and worms – been transformed through faults in some earlier step along the pathway to manhood that override its effects. One of the guilty parties resembles a sex-determining element found in worms. It sits on our chromosome nine – whose ancestor also gave rise to the sex chromosomes in birds. It puts a stop to manhood if one of its two copies has been destroyed by mutation. In this, too, we resemble birds, whose males also need two copies of the structure.

The gene involved – like *SRY* itself – makes a protein with a special segment able to bind to DNA. From worms to birds and people, the molecule is hard at work in the earliest stages of sexual development, well before the more obvious mechanisms kick into action. Manhood needs two functional copies of that ancient gene and, without them, SRY is impotent. In birds the equivalent DNA sequence is on the 'X' chromosome – found in double copy in males, but with just a single dose in their opposite numbers. Turtles and crocodiles, too, hint at its power, for in reptiles the hidden emperor of sex changes its activity with temperature.

Males, that mysterious element shows, have a history that stretches back at least to the split between worms and mammals a billion years ago. On that ancient foundation is balanced the gothic complexity of today, with one switch replacing another in different groups as the millennia roll by. In their urge to escape the attentions of their partners, females have obliged males to redefine the very mechanism that makes them what they are.

How did the endless race between the sexes begin? Some ideas about men are positive (even if few are as much so as those of the ancient Egyptians) and see them as a force for good, helping evolution on its way as they move genes between female lines. The evidence of universal conflict supports a grimmer view: that the parasitic habits of one party began as soon as the starting gun was fired.

Life managed without males for its first billion years, much of which was passed as single cells in a series of warm ponds. Then, in some ancient and neutral Eden, the fruit of the tree of sexual knowledge – a new mutation – persuaded members of a particular clone to fuse with cells from another, and then to divide. That ingenious idea is good news for the novel gene, as it doubles its rate of spread, but is a lot less so for those who receive it, who are obliged to copy the extra DNA. At once, two factions emerge, one keen to force itself upon the other. Thus was sex invented.

Soon one contestant began to cheat. Large cells are expensive, but are better at dividing because they have more food reserves. Small cells are cheaper to make, but cannot afford to split. Their sole chance of success hence lies in fusion with a large cell. The first males had appeared on the scene.

Their state fed on itself. For the new small cells the pressure to fuse is intense because they have no other hope of a genetic future. The more minute and more mobile they become, the more can be made and the better are the prospects of finding a large partner. Any that take extremism too far and become too small to swim fail. Natural selection pushes in opposite directions to make such structures tiny, or (as in females) huge. This inbuilt instability explains why there are only two – rather than dozens – of sexes; the male is reduced as far as is physically possible, and forces everyone else to put in whatever is needed for fusion.

The tension between sperm and egg and between those who make them has gone on for at least two billion years. It has driven much of evolution. For one of the parties involved, the first fusion marked nature's greatest mistake. For males, on the other hand, it was a triumph.

CHAPTER 2

THE COMMON MAN

'It is better', says the Chinese proverb, 'to raise geese than girls.' Geese, after all, lay eggs while daughters are nothing but an expense. Even to get rid of them a dowry must be paid. In China, at the time of the Cultural Revolution, the ratio of the sexes at birth was the same as that in the West, with about 106 boy babies born to each one hundred girls. Since then, the one-child policy has led to a great shift, with a 20 per cent excess of sons. Much of this comes from abortion and infanticide (and even parents who disapprove of such means are more likely to have another child if their first is a daughter; which alters their own prospects but not those of the population as a whole.

When it comes to animals, farmers often prefer females. Dairy farms have no need for bullocks, but are stuck with them. As a result, they are forced to turn half their calves into veal, with little profit. Now science has shifted the balance. Bull sperm are marked with dyes that bind to the X or the Y chromosome. The cells are then shot through an instrument – a fluorescence-activated cell-sorting or FACS machine – which diverts those with each label into a separate vial. It can manage twelve million sperm an hour, and in dairy herds the unwanted Y-bearers are thrown away. Before sperm selection, half a million bull calves from British farms suffered a subsidised death each year, but since the first three sperm-sorted cows,

Charity, Clover and Chloe, were born in 1999, their chromosomes alone have paid the price. The economics of milk production has been transformed.

Given the choice, Chinese and cattle-breeders each go for the more profitable sex. If their income depends on the sale of milk, daughters are better; but if the family's assets are safer in the hands of boys then it pays to invest in sons. Sometimes, cattle themselves are the means of exchange. To marry off their youths, the Nuer people of the Nile Basin must give animals to the bride's kin – but the bill is not paid until a son is born. Even a dead man can have male children as his brother sleeps with the widow in his name and, if successful, hands over the dowry.

Such an undue desire for sons seems to make no biological sense, for males are defined by the vast excess of small sex cells that they produce. Each makes more than enough sperm to fertilise a host of females. No daughter, in contrast, can have more than a few offspring. Why are those who make sperm so common, with the ratio of the sexes close to even? Why not just one lucky man and a horde of women?

Males began – as did sex itself – with economics; with the discovery by one player in the sexual game that it could subvert the investment strategy of the other. The key to their abundance also involves fiscal calculus: to transmit a biological heritage most effectively parents should put the same share of their resources into the passage of genes through each sex. Males are, as a result, much more frequent than might at first seem necessary.

One of the more improbable Utopias to decorate American fiction is Charlotte Gilman's *Herland* of 1915. Three explorers learn that 'the savages had a story about a strange and terrible Woman Land in the high distance . . . It was dangerous, deadly, they said, for any man to go there'. Intrigued, the young men enter. They are amazed at what they find: 'But this is a civilized country!' one exclaims. 'There must be men!' There were, in fact, none. All had been killed in a war long ago and parthenogenesis had taken over (the details are left obscure). The women of Herland learn from the intruders about the sexual world outside, with its murder, war and chaos and, after a series of feminist sermons, the young hopefuls are thrown out.

In any more realistic Herland – a society based on sexual reproduction and short of men – those timely trespassers would each find himself with a vast number of potential mates. Each might fertilise a hundred women. As a result, every owner of a Y chromosome, precious commodity as it is, would pass a hundred times more genetic information to the next generation than could any female. A gene for masculinity will, in such a Utopian state of affairs, be at a great advantage.

Any male in an unduly feminine place will be a huge success and his joyous state will continue until his fellows make up half the population. Then their advantages disappear. Should their numbers rise above half, the pendulum swings in the opposite direction as women become the more desirable commodity. It then pays to have daughters until, once again, the happy medium is restored.

The balance of the sexes is a matter of investment policy, in which the product in more demand commands a higher value. As in the world of oil, hogs or dairy futures, changes on the supply side can shift the balance. After the Hundred Years War, which killed vast numbers of soldiers, polygamy was legalised in the German states – but within a generation, as their numbers recovered, husbands were again limited to a solitary wife.

The economics of manhood is rather like that of agriculture. The Herlanders are shocked to learn of 'the process which robs the cow of her calf, and the calf of its true food; and the talk led us into a further discussion of the meat business. They heard it out, looking very white, and presently begged to be excused'. The laws of nature are, as those noble women feared, as merciless as those of the slaughterhouse. What counts is the survival not of children, but of genes.

An equal ratio emerges only when parents invest the same in sons and daughters. If it costs more to make one or the other, without an equivalent increase in the ability to pass on DNA, that class will become rarer. The expenses come before the young reproduce, the benefits not until they have done so; but the equation is so finely poised that its balance can change in an instant.

Parents can be persuaded to put vast amounts into their young as long as enough raw material is available. Today's champion milkers produce, with careful feeding, thirty thousand litres a year, far more than wild cattle invest in their calves. Bulls, too, can sire tens of thousands when the costs of courtship and copulation are driven down with a simple erogenous device. When life gets hard and the market takes a tumble, many parents are just as happy to take the simple economic decision to abandon, or to kill off, their young.

For selfish reasons a mother wants to see her progeny succeed in the coital battle, and speculates to improve their chances. A father has the same interests, but as he can slide off and find another partner he faces a constant temptation to reduce his outlay in each investment. Quite often, the divergent interests of each parent shift the balance of the sexes.

Each child is a gamble in DNA futures. If things look good, a high-risk bet is worthwhile; but if they are grim the time comes to retreat to a safer stock with a lower rate of return. In the animal world, a son in first-rate condition has the chance to attract lots of mates and to spread its seed to a large and appreciative market. Daughters are a better bet when times are hard as, however plain they might be, they will almost always find somebody desperate enough to mate with them and to transmit at least a few copies of their mother's heritage. A son in such a situation may have no chance at all.

A mother might as a result wish to put more into a Y-bearing fetus when the future is bright, and turn to the other option when prospects are gloomy.

The island of Rum, in the Hebrides, was once the preserve of aristocrats in the pointless pursuit of red deer. After a sudden financial chill they abandoned their Edwardian castle (a staring red folly with the first telephone system in Scotland) to biologists who pursue the same hobby for better reasons. The deer, they have found, defer to the economic rules that killed off the previous owners. A cold spring means poor food for mothers and a difficult start for their young. The sons of hungry mothers suffer the most, for they survive and reproduce less well than do their sisters. In wild mice, the male offspring of a starving mother do even worse, for they are reproductive failures for life, while their sisters thrive. The same is true for humans, as the underweight boys of mothers pregnant at the time of the Dutch famine in 1945 were less healthy and less likely to marry than were their girls.

For red deer and other creatures, the effect can last. A female deer calf underweight because her mother was short of food produces lightweight children herself, even when grazing is good, so that the trials of a parent are, like those of the family who once owned the island, visited unto the second generation, and perhaps beyond.

Birds play the market in other ways. They speculate not with cash or with food, but with chemicals. They deposit testosterone in their eggs, which – like ready money – gives a useful start in life. The more they put in, the more dominant a son. Zebra finch females given a superior mate, with bright plumage and loud song, invest with great enthusiasm. Those favoured sons grow faster, chirp louder and triumph in the erotic arena. A mother's choice alone is involved. An ugly partner can be made attractive with a bright red ring around his leg. Any bird imprudent enough to mate with this false Don Juan at once puts a dose of the hormone into her eggs (which costs her a lot, as it suppresses her own immune system) and her male chicks benefit.

In humans, property and the *SRY* gene also tend to travel together. Most nations see a father as the natural head of the household, but why sons are more valuable, nobody knows: perhaps men gained control because they needed to co-operate in the hunt, or because husbands with several wives wanted to keep other hands off their goods. Whatever the reason, paternal lines are the rule from China to Europe and fathers have been, for most of history, more important than mothers.

Patrilines – lineages of paternal descent – often define themselves by reference to an ancient progenitor. In 1819 William Moorcroft, Supervisor of the Stud for the East India Company's cavalry, met a Punjabi nobleman who could recite his own line back for four hundred and fifty generations. Other patrilines have little idea of who their common ancestor might have been and gain a sense of unity through the ownership (or at least the occupation) of a shared territory. Clans of this kind are now no more than curiosities, defined by somewhat eccentric gatherings of the Macdonalds, the Campbells and the rest in the Highlands and elsewhere. Once, they meant much more, for they gave sons political, as well as financial, power.

Before Romulus and Remus, the peoples of Italy were a collection of tribes in constant conflict. As the years went by, they united until three hundred patrilines, each with a chief in the Senate, formed the core of a new Empire. Daughters moved, while sons stayed to inherit the parental home. All clan members shared a grave (a habit still around today) and, at least in the early days, the paterfamilias had the power of life and death. In time, an alliance of fathers grew into a ruling class under an Emperor, the *pater* of the entire Roman *familias*. The system was designed to keep wealth within a noble line; and it worked.

Romans were serious about the link between cash and chromosomes. Daughters could inherit nothing, and any family without a son was finished. The law was strict: no woman could have proper heirs, and not until Christianity were patrilines separated from their property (much of which went to the Church). As man's economic power declined, wives regained some value.

In the natural world mothers, rather than fathers, have more control over their offspring's heritage. They can manipulate their legacy in many ways. Sometimes a mother decides which sperm are allowed to fertilise the egg, or puts paid to a fetus as it develops. As an expert investment analyst, she may feed sons, or starve them (or, when times are very bad, even eat them); but however she does it her decisions can cause great swings in the balance of the sexes.

In the lesser black-backed gull, a bird common on British coasts, sons are larger than daughters, which makes them more expensive to feed. A mother forced by unkind scientists to produce egg after egg by having each one removed as it is laid soon suffers. As her health declines, she manipulates her internal economy to produce more female chicks, sometimes by a ratio of three to one. Give her a series of good meals (boiled eggs, as it happens) and she shifts back to sons, the more costly option.

Wasps determine the sex of their young in a simple way in which the mother has complete control. She may choose to lay fertilised eggs that become females, or unfertilised versions that turn into males. Certain parasitic kinds lay their eggs in caterpillars, usually one per victim. The mothers are skilled gamblers. Faced with a small quarry, they insert a new-laid son, but with a larger prey turn at once to daughters (who gain more in reproductive terms from a generous diet as a larger animal can lay more eggs). Sometimes the pressures of mating itself change the odds. When the young of certain wasps hatch, they copulate within the host's body, brother with sister, and eat it alive. The first male to emerge has enough sperm to fecundate all his kin and does not hesitate to use it, so that the donations of his brothers are wasted. It then pays a mother to invest more in daughters, and the proportion of sons drops to less than one in ten. As a twist to the tale, if a second wasp finds the unhappy victim, she lays eggs primed to produce more sons – who have a chance to mate not just with their sisters but with the mass of unrelated females within.

Many biologists have claimed to find shifts in the sex ratio at birth among mammals, but these are small and inconsistent, if they exist at all. Life, like Wall Street, does not always behave as the economic rules say it should. Rhesus monkey mothers at the top of the copulatory tree have more sons, but high-rank hamadryas baboons have more daughters. As stockbrokers know, it is easy to make up explanations of such anomalies, but most are no more than guesses.

The tie between capital and copulatory success is also far from clear in human societies. Darwin, in sympathy with his eugenical cousin Francis Galton, was concerned that 'the very poor and reckless, who are often degraded by vice, almost invariably marry early, whilst the careful and frugal, who are generally otherwise virtuous, marry late'. There are few rules, but most of the figures are against him. Men who inherit wealth – like birds given a prenatal dose of testosterone – tend to spread their genes with more success than others. In nineteenth-century Sweden, three-quarters of all lower-class youths stayed single, compared to a fifth of the bourgeoisie. In ancient Rome, too, the urge to hand on wealth to a son turned on the vast improvement he made to the lineage's procreative prospects. The legal maxim was *Mulier autem familiae suae et caput et finis est*: 'a girl is the end of the family'.

Captains of industry and of empire may favour sons, but other parasites on nature's body politic disagree. For them, such creatures are a dead end. To escape from it, they subvert the economic rules of parenthood, to the great detriment of male children. From the point of view of life's hangers-on, the future is female.

Once, in the days when I did science, I spent a summer in the Rocky Mountains in the useful search for the highest fruit fly ever collected. I succeeded in my modest aim, with several splendid specimens collected at the tree line, eleven thousand feet above sea level. As fly collectors do, I set up a series of vials, each with a pregnant wild fly, and waited for their occupants to lay eggs. To my dismay, most of the cultures died out in the next generation, as no males emerged. At the time their absence seemed a malign trick of nature, but in fact it was an introduction to a hidden world of conflict which much distorts the balance of the sexes.

Males are, in the genetical sense, parasites upon females. Selfish as they are, they pay a terrible price for their habits, for they face a set of real parasites who depend on the second sex for their survival.

Females have large reproductive cells, which pass on not just a mother's DNA but a mass of material useful to the embryo. In addition, their eggs are filled with hangers-on who hope to hitch a ride to the next generation. Many are harmful, but some have become part of the body's machinery. Mitochondria are the structures within which every cell burns its chemical fuel. Crucial though they now are, they were once bacteria (and their closest relatives are still the cause of typhus). Certain intracellular passengers still take the trip without paying for the privilege.

Their genes, like all others, have but a single interest: to copy themselves. They can do so through eggs alone, as sperm are stripped-down objects that pass on only the DNA of the individual who made them. As a result, any fellow traveller unlucky enough to get into a son – which, given the laws of genetics, happens half the time – is doomed. It will leave no descendants, unless it can change the rules. For cellular colonists, helpful or otherwise, males are the end of the line.

If sperm is a dead end, the best strategy is to do away with it. The host – needless to say – does not much like the idea, as it transmits its DNA most efficiently by investing the same effort into sons and daughters and does not want to abandon one of them. As a result, economic warfare breaks out between invader and invaded. Quite often, the parasites lurking in the cytoplasm succeed. They have made a strange and widespread series of attacks on males. Life has come up with many inventive ways in which to abandon sons, and the tiny structures within the egg have tried most of them.

One of the creatures responsible is a bacterium called *Wolbachia*. It is the Herod of the bacterial world; a great killer of sons, first-born or otherwise. It spends its life within a variety of insects, mites, pill bugs and worms. Thousands of species of insects suffer its advances and a single fruit fly may bear millions of the unwanted visitors. As is the case for many parasites, *Wolbachia* has a tiny genome and, like some other agents of disease, the tree of relationship of the various kinds differs from that of their hosts, as evidence that they often hop between species. Some *Wolbachia* act rather like infectious diseases (and the bacterium is related to pathogens such as the agent of heartwater, a tick-borne ailment of African cattle), but others have become essential to those who bear them.

Wolbachia does little harm until the time comes for copulation. Then the ingenious but irksome beast makes its presence felt. As eggs alone pass it on, the makers of sperm suffer the lash of bacterial disapproval, inflicted in various ways.

In legend, Iphis' mother was ordered to kill her child, if it were born a girl. It was, but she brought her daughter up as a boy. Her father then betrothed his false son to Ianthe and – luckily for both parties – a convenient god changed the child's identity in time. The Nuer manage without divine help. A sterile woman defines herself as a man and persuades another female to marry her. She chooses a male to sleep with her new wife. His task complete, he wanders off, and the children then refer to her as their father. A quick gender reversal allows an unfruitful lineage to survive. *Wolbachia* does the same, more brutally, and in reverse.

Sometimes it behaves like a venereal disease. If an infected male mates with an uninfected partner, the eggs die, while crosses in the opposite direction – infected female with innocent provider of sperm – succeed. Somehow the bacteria arrange matters to ensure that only the eggs that contain copies of their kin succeed. No uninfected animals are born (which eases the pressure on those who have picked up the hitchhikers) and the parasites survive while half their hosts suffer. The intruders may go further to help themselves. In ladybirds, infected sons die when they are still eggs, which gives their sisters a start in life as they feed on their brothers' corpses.

For some animals Herland is reborn ('One of these young women bore a child. Of course they all thought there must be a man somewhere, but none was found. Then they decided it must be a direct gift from the gods . . . This wonder-woman bore child after child . . . all girls'). In woodlice, *Wolbachia* achieves its Utopia by switching off the gland responsible for masculine hormones, so that only females are born.

The unfortunate male may be disposed of in a more ingenious way. Bees and wasps lay fertilised eggs which become daughters, or unfertilised eggs which develop into sons. *Wolbachia* manipulates the system to persuade its hosts to have daughters alone as it can force eggs to receive a double copy of the mother's chromosome set – which obliges them to develop with that identity even though no sperm is involved. The tactic is ancient indeed, for some tiny parasitic wasps in which males have never been found suddenly produce them after a quick course of antibiotics, proof that they had been suppressed by the feminists within.

So few producers of sperm may be left after a *Wolbachia*

attack that their partners are forced to abandon their normal coy behaviour. A certain species of African butterfly is in some places reduced to but a single male in twenty. Most females stay virgins. In a desperate attempt to find a mate hundreds fight and chase each other in great clouds around the local hilltops. Any male lucky enough to stray into this lepidopteran Herland copulates at once. The shift in proportion makes large sex cells cheap and small ones valuable, and leads at once to a remarkable role reversal.

If the bacterial assault is too successful, it might get rid of sons altogether, which would mean disaster for both host and parasite. Real life is more complicated and allows both parties to survive. Uninfected animals fly in and insects attacked by the bacteria are sometimes cured of their bias by natural antibiotics or even by hot weather, both of which kill the invader.

Like all diseases, *Wolbachia* copies its own genes at the expense of those of its host. At once, the pressure is on to fight back. Genes in the nucleus are not content to accept the loss of half their number. As a result, many devices have evolved to suppress the action of the internal man-haters. Sometimes the battle ends in truce and the parasites become useful to their hosts. As a result, certain creatures – such as the agents of elephantiasis and river blindness in Africa – die after a dose of anti-*Wolbachia* drugs (which is a beautiful example of un-expected science: who would have dreamed that a dose of tetracycline might cure a disease caused by worms, who are normally quite unperturbed by such medicines?).

The unfortunate male suffers the attacks of sexual foes who come from outside. He must in addition cope with traitors; with the enemy within, who cheat on the molecular market to bolster their own prospects. Again and again, animals suffer mutations that subvert the biological rules and spread at the expense of their carriers, but again and again such self-centred genes are held at bay.

Even mitochondria can slip up to reveal their dubious ancestry. Any foolish enough to enter a sperm are doomed because they are destroyed when they enter the egg. In hermaphrodite plants, from beans to maize, the mitochondria save themselves with mutations able to sterilise the male organs and to force the plant to make eggs, their vehicle into the future. The nuclear genes can bite back, with matching genetic changes that suppress such attempts to geld their carriers.

My montane and female fruit flies, in contrast, carried an altered and dishonest X chromosome. It causes Y-bearing sperm to be eliminated, so that fathers who inherit the selfish X produce only daughters – all of whom bear the guilty DNA. Such genes are common in flies from the southern half of the United States and from Mexico, but are absent from the north of the continent. Why, nobody knows, but some rival force must limit their spread.

The Y itself may lead the fight back. In certain fruit flies, it carries genes able to rescue those who bear a selfish X from their fate. How well it succeeds varies from species to species, and from place to place. Some populations have lots of devious X chromosomes but no protective Ys, while others, in rather an inscrutable way, show the opposite pattern. As so often in the world of reproduction, the two players are in an endless arms race, with sometimes one partner ahead and sometimes the other. The battle of the chromosomes is of great interest to genetic engineers, who may soon be able to shift the relevant genes to domestic animals, to get rid of bulls for dairy farmers and cows for breeders of beef, with no need for complicated equipment.

And what of ourselves? Has evolution done to *Homo sapiens* what it did to flies or beans? Sons are certainly expensive to those who bear them, as mothers take longer to have a second child if their first is a boy. In pre-industrial times, as manifest in the records of nineteenth-century Lapp hunters, every son reduced his mother's life expectancy by six months or so, while a daughter slightly increased it. The simple pleasures of maleness do lead to a shift in the sex ratio as the years go on, as they kill off men faster than women. But is there any natural variation in the balance at birth?

The evidence is at best equivocal. Part of the problem is that families with unusual proportions of one or the other come to attention while others do not. In one case, seventy-six daughters were born over three generations in a particular lineage, with no sons at all – but with millions of families in the records such figures could emerge by chance alone. Some theories seem doubtful – does the balance really change after earthquakes or among the children of deep-sea divers? – and there is a fatal tendency to comb through vast bodies of data in the search for statistical flukes. At best, such effects are small, unreliable and hard to understand.

Even so, we do show small shifts in the numbers of sons and daughters, in which genes might play a part. Families already blessed with sons are more likely than average to have another, while the opposite is true for those with lots of daughters. Fathers are not to blame, because they make the same number of Y- and X-bearing sperm. Neither does the birth of a boy change a mother in some way to make her favour male pregnancies. Instead, to a tiny degree, some mothers have more boys while others bias their broods in the opposite way.

Darwin himself tabulated the British figures, and found them to be almost the same as today (although the Welsh had fewer girls than expected). In Europe the proportion of boys rose after the First World War, but then fell, only to edge upwards for a time after the second global conflict. Older parents have more girls, and a change in the average age of parents might perhaps explain the shift. For some reason, marriages of parents of different race also give rise to more boys. Tobacco kills smokers and may do the same to sperm. Those who smoke more than twenty cigarettes a day have fewer boys than average. The effect is stronger for fathers than for mothers – perhaps because sperm that bear a Y chromosome, fragile as it is, are more susceptible to the poison.

Affluence may also ruffle the marriage bed. The hundred and fifty children of US Presidents since Lincoln show a great shortage of daughters, and the Britons who make it into *Who's Who* face the same problem. Some claim that female trial lawyers, an aggressive and successful group, have more sons, while beauty queens (the epitome, perhaps, of lady-like passivity) go for the alternative. Although many have searched for it, the West today has no general fit between social class and the births of boys and girls.

By far the most important cause of a shift in the balance of the sexes is large, simple, brutal and hidden from the eyes of seekers after the truth. It has been around for years and is discussed at length in *The Descent of Man*. As the Roman proverb put it, 'the poor have no daughters'. In much of the world the same is true today. Millions of girls are destroyed at or before birth, as testimony to their worthless state.

In the West, young boys die at a higher rate than their sisters. In two-thirds of the less developed world, the opposite is true. The murder of female children is common and, in the new global economy, has become more so.

An India freed from imperialism would, said Nehru on Independence Day, build up a prosperous, democratic and progressive nation, and ensure justice to every man and woman. For men, his nation has done well, but their wives and daughters have been less fortunate. Half a century on, they are, in comparison with their partners, less involved in the economy, have lower levels of literacy, and experience a larger gap in life expectancy than they did in the days of the Empire.

The murder of girls is a valued Indian tradition. Rajputs, Sikhs and other warrior castes preferred to marry their daughters to a husband of higher rank – which meant an expensive dowry, or the rapid disposal of the unwanted child at birth. The British became concerned when they saw the results of the first census of 1871. In some villages, the commissioners reported, not a single female child was to be found. The authorities brought in the Female Infanticide Act, which set heavy penalties on child murder, and policemen were stationed in such places – but, twenty years later, some provinces still had twice as many boys as girls.

For a time, the habit began to fade, but now things have changed for the worse. Dowries often take half of a poor family's disposable wealth and the death of unwanted children has become more, not less, common with India's new affluence.

The value of females depends on the market. In Kerala, a liberal society with an educated population, daughters are born unscathed. In the north and northwest of the country, in contrast, tradition rules. All over the Punjab, Haryana and the United Provinces, men want large families, and many children die young. Women move out of the household to marry, and are rarely seen in public once they have done so. The economy is based on wheat, rather than – as in Kerala – on rice. Wives play almost no part in the fields, and their value has been further reduced by the farm machinery brought in after the green revolution. Girls suffer as a result (the sole exception lies among the untouchables, whose poverty is such that the efforts of all children are needed to keep the family alive).

In certain villages, young boys outnumber girls by three to one. Week-old girls die at twice the rate of their brothers. Often, neither the birth nor the death is recorded; but, when parents admit their child's demise, the cause is given as 'pneumonia', or that 'the baby became stiff'. Boys, the villagers tell the curious, are saved from such a fate because the correct gifts have been made to the gods.

Mothers stay in hospital for several days when blessed with a son, but after the birth of a daughter leave at once. *Dais*, traditional birth attendants, often kill the child, for a fee of around a hundred and fifty rupees (about two pounds sterling). They can, they claim, assess a baby's gender even before birth, and stand ready to do their duty. In some places, each admits to a murder a week. The relatives may do the job themselves by forcing the mother to place tobacco under her child's tongue. If she refuses, she is herself killed or thrown out of the house.

The practice was once limited to the higher castes, but a desire to copy their betters, combined with economic pressure, has spread it even to Sikhs and Christians. As the Indian economy has evolved, so have the reproductive rules. The Kallar community in southern India was branded a criminal tribe in the days of the Raj and many of its members were imprisoned for banditry. Their women were assertive, worked hard and supported their kin while their husbands were out of circulation. Their villages were poverty-stricken, but both dowries and infanticide were unknown.

In 1958 a dam was built. A few communities could grow cash crops and became wealthy, but most stayed poor. At once a dowry system began as parents became desperate to marry their daughters into a richer household. Now the incidence of child murder is among the highest in India. In one recent year, five hundred and seventy of the six hundred girls born were dead within days. So scandalous were the figures that the law at last became involved. For the first time in India, somebody was found guilty of child murder and went to prison. She was, needless to say, a woman – but who was really responsible? No man was charged with any crime.

Nowadays a husband's relatives ask not for clothes, but for televisions; and, in the slums of Bombay, a dowry may represent five years' worth of household expenses. The Dowry Prohibition Act of 1961 has had no effect. Twenty years on, only the rural northwest had much of an excess of boys soon after birth. Ten years later child-killing had spread, and for the first time in history, India's cities now as a whole have a masculine bias. A dreadful recent development involves dowry murder – young brides whose families have not come up with the goods are burned alive, often with the pretext of an accident with a kerosene stove. At least two thousand wives a year are the victims of such crimes, which did not become common until the 1970s.

The neglect of females is as old as civilisation, but science has made the problem worse. Now, it is easy to check the fetus – and those without a Y chromosome suffer. The World Health Organization insists that sex is not a disease and the idea of pregnancy termination on the grounds of gender alone (except when necessary for medical reasons) is banned in Europe.

India has a different view. Most citizens believe that prenatal diagnosis was in fact invented to check the gender of a fetus rather than to help with genetic disease. 'Select' (a herbal remedy which, said its makers, turned unborn girls into boys) has been banned, but science has taken its place. Prenatal tests of this kind were forbidden in 1996, with a three-year sentence and a heavy fine, but the law applied only to government health centres and not to private clinics (even if the frank slogan 'Better 500 rupees today than 500,000 tomorrow' has been quashed). Bombay alone has two hundred clinics that offer such screening, with almost all the abortions aimed at daughters.

Now, the job has become easier, with portable scanners taken from village to village to check whether a fetus passes its prenatal examination. Up to a million unborn girls are destroyed in India each year. To check whether his wife is pregnant with the wrong kind of child costs an unskilled worker two months' wages, but the fiscal balance makes it worthwhile. For the middle class, private clinics have begun to provide test-tube fertilisation followed by selection of the desired type; and, of the few dozen who have so far used their expertise, every one has asked for a boy.

The murder of children which so shocked the British in 1871 led to a ratio of nine hundred and seventy-two women to a thousand men. In 1991 modern science had shifted the figure to nine hundred and twenty-nine females to each thousand males. Gandhi's goal of a nation in which, intellectually, mentally and spiritually, women would be equivalent to men has not been realised.

Whatever the social pressures to get rid of one sex, the simple laws of reproduction will, in the end, always make their presence felt. To interfere with them can lead to painful – and expensive – consequences.

In China girls are still sometimes called 'Too Many' or 'Little Mistake' to reflect their value, but once they were worth even less. In the nineteenth century the province of Huai-Pei suffered a series of famines which led to civil war. Daughters were despised as another mouth to feed and, quite soon, their numbers began to plunge. As their brothers grew up, they found nobody to marry. Great gangs of disaffected youths grew into a horde of a hundred thousand rebels – the Nian. They almost overthrew the Imperial dynasty before they were crushed.

The problem of the friendless and discontented Chinese youth has returned. Even official statistics (which understate the problem) suggest a ratio of a hundred and seventeen male births for each hundred females – which gives the nation eighty million young men with no hope of marriage. The age gap between groom and bride has increased as older men take teenage wives (which makes life even worse for the next generation), and bachelor villages have appeared in distant provinces. The residents of such places ('bare branches', as they are known) once became monks, or soldiers, or eunuchs in the royal household. Now they move to the cities and add to social unrest. There has been an outbreak of abduction of girls, who are sold into families in search of a daughter-in-law or as prostitutes in the male-filled cities.

The government sees the problem. Selective abortion of daughters has been banned, and posters proclaim that 'Girls are fine descendants too'. It will, alas, take more than slogans to remove a habit built so deep into the nation's fabric.

The developed world, less bound by economic pressure, is not much concerned with gender balance. In a few places in nineteenth-century Germany younger sons were killed at birth, supposedly to avoid dividing the family estate in the next generation, but today's opinion polls suggest no overall preference for sons or for daughters (although individual families may desire one or the other). There is not even much evidence that boys and girls are treated in a different way when it comes to inheritance. In Canada the rich leave more to sons and the poor to daughters, but in California the opposite is true. Every claim that the affluent favour male children is countered by another that finds the opposite. A vast survey of American attitudes, from mothers on welfare to millionaires, showed no consistent preference for investing money – or even breast-milk – in one sex over the other. In the developed world people are good, or bad, to children without much reference to whether they bear the *SRY* gene.

Even so, families tend to stop as soon as both boys and girls have been born. Parents with two boys, or two girls, choose to have a third child more often than do those whose first pair were of different gender. This affects their own household, but has no effect on the overall balance.

Genetics stands ready to help, if asked. In the United States a third of counsellors would allow a pregnancy termination for a couple who want a child of a particular gender even if no medical issue is involved. The figure for Israel is twice as high. Britain has so far been strict, except to avoid inborn disease. In 1999 the Mastertons of Edinburgh lost their three-year-old Nicole in a fire. With four sons, they were desperate to replace their beloved daughter. The authorities refused even to allow them to choose to implant an egg fertilised by an X- rather than a Y-bearing sperm.

Plenty of couples try to subvert the rules of nature in a less drastic way and are happy to pay for the privilege. The author of *How to Choose the Sex of Your Baby* retired to Las Vegas on the proceeds. His method turns on the unsupported claim that sperm with a Y swim faster than do those with an X. To have a son, he claims, hopeful mothers should copulate on the day of ovulation and try to have an orgasm. The technique works no better than those which depend on the phases of the Moon (or on the husband hanging his underpants on the right rather than the left of the bed), but half his customers were satisfied and his pension was assured.

The FACS machine, used to sort cattle sperm, has now been turned to our own ends. The businesses who shift the ratio for cows have been joined by the MicroSort Company, which does the same for humans. It helps those who look for what its directors call 'family balancing' (as the American Society for Reproductive Medicine puts it, 'sex selection might provide perceived individual and social goods such as gender balance or distribution in a family with more than one child, parental companionship with a child of one's own gender, and a preferred gender order among one's children'). So far its services are restricted to couples who already have two or more children of the same sex – and who are happy to be counselled to ensure that they do not reject a child of the unwanted variety should the device fail.

The machine binds a coloured dye onto the X or the Y. It has a success rate of about nine in ten for those who wish for a girl, and (given the smaller size of the Y chromosome) seven in ten for people with the other inclination. At two thousand dollars a try the procedure is not cheap, but already five hundred or so American pregnancies have come from sorted sperm. In stark contrast to the parents of the developing world, Americans who make such choices much prefer girls. Three-quarters of the clients ask for a daughter rather than a son.

In Herland itself, that 'strange and terrible Woman Land in the high distance', one of the explorers falls for a local girl and tells her of the biological advantages of sex. She becomes convinced of the superiority of a world with men in it and moves to the United States with her new husband. In that land of sexual Utopia, her sisters and daughters seem safe. America might, in time, with the help of MicroSort, even evolve into some version of her homeland, with an excess of women, the more valued gender.

The global shortage of a hundred million of their fellows is a reminder that in other places the economic sums add up in a different way. Remorseless as such calculations may be, human evolution follows the rules of other creatures and the laws of nature are likely to win in the end. When it comes to men, popularity does not always mean success; and from Herland to Hindustan – and whatever man's preferences – the balance of the sexes will some day be restored.

CHAPTER 3

SEVEN AGES OF MANHOOD

Russian politics since the Revolution has been driven by testosterone. The reins of power passed from Lenin (bald) to Stalin (hairy); to Khrushchev (bald) to Brezhnev (hairy) and in rapid succession to Andropov (bald), Chernenko (hairy), Gorbachev (bald), Yeltsin (hairy) and then to Putin, the current President, who is noticeably thin on top. The depilated were, some say, reformers while those with hair were reactionary. On those grounds (and not all the evidence is dependable) US senators are a conservative lot, for they retain more cranial coverage than most. The American trichological cycle takes longer than the Russian. Several Presidents before Eisenhower were bald, but nobody with the condition has been elected since the 1950s (and in every election bar one since then, the best mane won). The fate of William Hague – and, no doubt, of Iain Duncan Smith, his successor as leader of the British Conservative Party – shows that the rule is now universal.

Bald heads – like beards – are secondary sexual characters, more frequent in men than in their partners, but not directly involved in the mechanics of reproduction. Darwin devoted many pages of *The Descent* to man's hairiness. He blamed sexual selection for our naked skins and felt that 'man, or rather primarily woman, became divested of hair for ornamental purposes'.

However they evolved, beards and bald pates are mile-stones in the seven ages of masculinity. After a sexless infancy, the embryo can for a time develop in either direction. It then creeps unwillingly to accept the destiny set in its DNA, although it may still be diverted to a different end. At birth, act three in the lifelong drama, boys and girls are quite distinct. Not until puberty does testosterone, the prince (and princess, for it is important in females) of hormones, cast its spell on bodies, pates and minds. In the next, fifth, age – the adult years – husbands and wives diverge, each with their own set of internal rhythms. Then, for one party, comes the menopause: a hormonal shift shared, to some degree, by their mates. History ends in a seventh and post-sexual world, a second childishness in which the players lose their differences in the face of oblivion. Each step in the male career is marked by a change in hormones and in their proxy, hair.

Each age is an exit and an entrance and each puts a label on those who pass through it. Perhaps that is why baldness is so unwelcome. All suffer in the end, but those who lose their thatch when young send out, too soon, a message of decay. Just one hairy male in ten would mourn the loss of his cover, but most of those who suffer it are filled with concern. Their worries are justified, for in a survey sponsored by a wig company girls found just one in six pictures of potential partners with naked scalps attractive, compared to half of those not so afflicted. Few are as sensitive as Elisha ('little children . . . mocked him, and said unto him, Go up, thou bald head; go up, thou bald head. And he turned back, and looked on them, and cursed them in the name of the Lord. And there came forth two she-bears out of the wood, and tare forty and two children of them') but the Bald Headed Men of America from their head office in Morehead City, North Carolina, estimate that a billion and a half dollars is spent each year on hair replacement.

The cause of the condition has been much debated. In 1896 *Scientific American* found brass instruments to be 'deplorable', with the trombone and French horn quite fatal. In the army, indeed, the condition was known as 'trumpet baldness'. 'The piano and the violin, especially the piano', on the other hand, had an 'undoubted preserving influence'.

Hair, in its loss and gain, is masculinity in miniature, from conception to coffin. Where and when it grows depends on commands from the Y chromosome, the testis and the brain. Their many errors in transmission, reception and translation hint at the complexity of the male machine.

At least in outline, the directions on the road to adulthood – hair included – are simple. The default is to be female and without an effort it is hard to become anything else. In 1889 the French biologist Charles Edouard Brown-Séquard, then over seventy, declared his youth to be restored. He had injected himself with a *liquide testiculaire* taken from dogs, and felt stronger and less constipated than before. Even the arc of his urine stream increased by several inches. Soon the magic fluid was used to treat cancer, gangrene and hysteria. His claim was illusion, but Brown-Séquard's testicular fluid revealed a great truth: that manhood is chemistry.

A hint at the nature of the reaction came when female calves with a male twin were found to be masculinised 'freemartins' – but only if they shared their brothers' circulation before birth. Then came a crucial experiment. Rabbits castrated as embryos grew up as females, but could be rescued from their fate with an extract of testis – proof that sex was not fixed, but was controlled by something in the blood.

The guilty party was hard to chase down. In the 1930s German scientists distilled six thousand gallons of policemen's urine in a fruitless attempt to isolate their virile essence. Undeterred, they then ground up a ton of bulls' testicles, with equal lack of success. A less ambitious group of Dutch researchers went on to macerate the testes of uncounted mice and hit the jackpot. Testosterone had been found. The substance is but one of many hormones that mark the path from Y chromosome to adulthood.

Such substances fall into three groups. The first is based on fatty materials, the second on proteins, and the third on a set of unrelated compounds. All the members of the fat-based group, testosterone included, derive from that infamous substance, cholesterol, and each is built on a carbon backbone which somewhat resembles a three-dimensional climbing frame. The hormones most concerned with masculinity – the androgens, as they are called – have nineteen carbons in the frame, one more than their relatives, those (such as oestrogen) more often associated with females. Their masters, the protein hormones, work closer to the centres of command in the brain, while the third group of emissaries ranges far and wide. Most of the substances fit into special receptors on target cells and act to switch particular genes on and off. The levels of such compounds rise and fall hour by hour, day by day, and year by year. Together, they link the web of sex to a variety of other bodily functions.

Most of the internal controls are not limited to one party alone. Testosterone does important tasks for women (the growth of pubic hair included). Oestrogen, too, is quite at home in their partners, where it is made from its relative by a special enzyme called aromatase. As a result, parts of the testis have more receptors for the hormone than does the ovary, and at times sperm is more exposed to it than is the egg. All men over fifty have more oestrogen than does any post-menopausal woman. Its name comes from the Greek for 'gadfly' and in men it lives up to its title. Those with aromatase problems have little interest in copulation, while mice unable to take the hormone into their cells court with enthusiasm until it comes to the moment of insertion, when they wander away. A few boys share the problem and they too are unable to mate. To balance the equation, and to hint at the role of male chemicals in the most feminine of events, females with a damaged testosterone receptor cannot give birth.

Man's protein hormones also play an important part in the menstrual cycle, as their names – follicle-stimulating and luteinising hormones (which in females control the release of eggs) and prolactin (involved in milk production) – suggest. The milk hormone increases rapidly in male marmosets (in which a father carries one of the twins borne by his mate) after their partner has given birth. It might also rise in human fathers as they bond with their baby.

The protein at the centre of command, gonadotrophin-releasing hormone, is made in the hypothalamus, the brain structure responsible for the status quo of temperature, weight and blood pressure. The hypothalamus rules the sexual empire and with the help of various secondary substances, from melatonin (which responds to day length and is important in seasonal breeders such as sheep) to endorphins (which increase contentment, pain tolerance and the like), links the head to the genitals.

Sex hormones switch genes on in the right place and at the right time and ensure that functions good only for one partner are not expressed in the other – which explains why genes for masculine attributes are not confined to the Y. Instead, they are scattered over the whole genome, but are turned on in men alone.

Their presence in their partners can be revealed with a quick dose of testosterone, which can cause hen birds to grow showy plumage, or ladies to grow a moustache. Men, too, hide some unexpected talents. *The Descent of Man* discusses the knotty question of nipples: as it points out, 'their essential identity in the two sexes is shewn by their occasional sympathetic enlargement in both during an attack of measles'. When the chemical balance is disturbed, they may leap into action. Some baby boys make 'witch's milk' under the influence of their mother's hormones before their own take over.

The same happened to prisoners of war in 1945 who, well fed after years of starvation, grew breasts and lactated until their bodies returned to normal. A certain gene important in sperm development is even involved in hereditary breast cancer.

Males may, when given the chance, reveal their feminine side, and females – drab as they often are – carry the information needed to make masculine traits. All this hints at the divergent interests of the two. Each limits, as much as it can, the harm done by a feature good for the other with a judicious dose of hormones, which forces genes to do their job only in the right place.

The body's ability to curb the malign effects of mutations that favour just one sex has caused the X chromosome itself to take control of a number of masculine functions. Many genes active in the sperm pathway live there rather than on what seems their natural home, the Y.

Their unexpected abode is a consequence of the Y's reduced state. Most chromosomes come – like kidneys or eyes – in double copy, and (as for kidney failure or eye injury) the effects of any damage are reduced by the presence of an unaltered version. Whenever it finds itself within a male, however, the X chromosome has no protector. Colour-blindness is a hundred times more common in boys than girls. The guilty gene sits, like thousands of others, on the X and, because its partner cannot mask it, always manifests itself when in a son. In daughters the damage is usually hidden.

Dangerous mutations on the X hence tend to disappear quite soon, but its pattern of inheritance also means that whenever the chromosome finds itself within a male, any mutation good for him is favoured. Even if such changes harm his opposite numbers, they spread for a while, for in them the damage is often concealed by an unaltered version of the chromosome. As the new gene becomes commoner, more and more females inherit two copies and suffer as a result.

The pressure is on to restrict the benefits to a single sex and to limit the damage done to the other. Hormones do the job, and, in time, the presence of genes for masculine function on a largely female structure is as well hidden as are those for a woman's ability to grow a moustache.

Hormone levels are under fine control. Some — testosterone included — respond to their own presence: too much, and the factory shuts down; too little, and it speeds up. Others are under the direct restraint of the brain. The prime tincture of maleness is itself kept in stock until needed. Testosterone is a hundred times more concentrated in the blood than at its target sites, but spends its time bound to proteins which do not release it until called upon.

Some enthusiasts, in a misguided attempt to become more manly, interfere with their own internal balance. They take anabolic steroids, drugs able to promote growth and repair which are used to help people with burns or those who suffer from osteoporosis. In the United States (but not in Europe) some versions are available over the counter. After cannabis and amphetamines they are the third most common drug offered to British schoolchildren. Fitness fanatics may take fifty times the recommended dose and in some gymnasiums one customer in ten admits to their use.

Half of all Americans are unhappy with their physical appearance, and many try to change it. Men in that predicament may turn up in a surgery with 'bigorexia', a condition in which massive muscles are matched with tiny testicles and an inability to make sperm. High levels of artificial steroids (and the United States has a billion-dollar black market in such things) have interfered with the production of their natural equivalent. Some abusers return to a sort of adolescence, with acne and a squeaky voice, while others enter premature old age (with baldness thrown in). A few grow breasts as the alien substances are converted to oestrogen. In an attempt to return to normal some turn to tamoxifen, a drug invented to block hormone receptors as a treatment for breast cancer, while others take drugs used to help restore fertility to women unable to bear children. Such attempts to escape from the athlete's dilemma almost always fail.

Androgens, natural or otherwise, bind to special locks within the cell that open to their molecular key. These androgen receptors, as they are known, are proteins a thousand or so amino acids long which stand ready to pass on the joyful news of an order from maleness headquarters to the genes poised to take advantage of it. They are part of a great family of biological ushers, each ready to welcome its chosen guest. The receptor has a segment that binds to the visitor and a hinge that allows it to change shape. Then it attaches itself and its emissary to special sequences of DNA close to the target gene. This switches the gene on and masculinity surges forward. Some receptors are present in all parts of the body (the brain included) and others in a few; some bind strongly and others scarcely at all; and some are faithful to a single messenger while others listen to many.

Most of man's finest attributes (penises and naked scalps included) need in addition a local enzyme which transforms testosterone into a more potent substance. In turn, that causes them to spring into life. The presence – and the size – of such traits hence depends on how much of a hormone, of its receptor and of the transforming enzyme are made, and when and where they act. A hierarchy of command, with various checks and balances, persuades certain organs to develop with one identity rather than the other and hair to grow – or fall out – in masculine places.

Freud's antique notion of women as diminished men is quite wrong. Biology reveals instead every man's battle to escape the woman within. From conception onwards, male chemistry pushes the embryo into a difficult and uncertain endeavour. Baldness may be a minor penalty, but it hints at the many risks taken by all those who travel towards that goal.

Hair grows from five million or so follicles, factories in the skin that squeeze a soft mass of a protein called keratin through a tight sheath. Each hair extends by six inches a year and, at the end of its natural life, falls out. The follicle rests for a few months, and starts again. The period of growth – five years on the scalp, three months in the eyebrows – determines how long any hair becomes. The brain plays its part; one bored lighthouse keeper weighed the contents of his electric razor each day and found that the amount doubled on Fridays, just before he returned to his fiancée on the mainland.

Lustful anticipation may cause hair to grow, but its loss cannot, alas, be curbed by pure thoughts. Most men lose half their follicles by the age of sixty but some exhaust them much earlier. Ten million Britons perceive themselves as bald. As the years go by their cover grows finer until at last it fails to surface at all. A tide of skin spreads until a mere reef of hair is left. For some unfortunates, even that sinks altogether.

Baldness runs in families. John Adams, the second US President, suffered; his son, the sixth holder of the office, had the same problem and his hairless kin are around today. The condition was once thought to be due to a gene able to manifest its effects in sons alone, but passed through both mothers and fathers. Just a single copy was said to be needed, so that half a bald father's sons were expected to be bald. The idea is too simple, as more than half the hairless have fathers with the problem. Several genes, rather than one, are probably responsible. A DNA probe is now available which will allow some young and still hirsute men at risk to discover, if they wish, whether or not their hair will fall out later in life.

Those with nude pates may be at more risk of heart disease than others, but the claim that such people are oversexed – Romans chanted politely to Julius Caesar: 'Guard your wives – the bald adulterer comes!' – is no more than myth. Caesar and his fellows do not have exceptional levels of testosterone but are instead cursed with an unwanted hot spot in the ability to respond to it. They have a surfeit of androgen receptors on the top of the head, together with lots of the enzyme able to transform the chemical into its more potent form. The substance rushes into the follicle and, in its excess, persuades the cells to give up their job. As a result, the remnants of hair on an otherwise naked skull contain three times as much of the hormone as normal.

Hair can fall out for several other reasons. Members of a certain Pakistani lineage who lack even eyebrows bear an inherited error in the ability to read a DNA message in the hair pathway. In mice, this *nude* mutation interferes with protective white blood cells and the animals get cancer as a result. Fortunately, the Pakistanis do not have this problem. Mouse mutants hint at the presence of other genes. Some damage a growth inhibitor and the animals sprout a long rich coat as a result. The Angora mouse has a cat to match, but not a human. Not all scalp problems come from faults in the DNA. The immune system can attack the follicles and some people lose their hair after a sudden stress, be it surgery, starvation or illness.

DNA determines not just how much hair grows, but where it decides to do so. The bald often have well-covered chests, and a rare mutation on the X chromosome causes a thick mat to grow all over the face. Although Darwin was much taken by the issue of why races such as his despised Fuegians are less hirsute than others, while Europeans have thick beards (which he saw as a 'primitive' character), such questions remain unanswered.

Baldness, the totem of masculinity, once seemed simple, but is not. In its complexity, the life and death of hair is a microcosm of male development. That, too, can fail in several ways. Its errors – tragic as they are to those who suffer them – help to understand what makes a man. They show how the path to that destination is full of pitfalls, and how the traveller is, as he trudges towards adulthood, in constant danger of being forced back onto the broad path to femininity.

Pierre de Coubertin, founder of the modern Olympic Games, saw women's sports as against the laws of nature. He followed the rules of the ancient Greeks. Wives were forbidden even to be spectators at the event (the penalty was to be thrown off the Typaion Cliff). In 440 BC the widow Kallipateira, who had trained her son as a boxer, disguised herself as a man to gain entry. She was overcome by emotion when her boy triumphed, and leapt into the ring. Her robe caught on the barrier and was torn off, to reveal her true nature. She was forgiven because of the exploits of her child, but from then on both trainers and athletes had to appear naked in the arena.

For the Greeks, as for de Coubertin, the very act of involvement in the Games defined a competitor as male. That Olympian view faded, and women were allowed to take part in 1900 – which at once suggested an ingenious fraud. In the 1936 event, Hermann Ratjen, a member of the Nazi Youth, entered the women's high-jump competition as 'Dora', to replace a Jewish contestant banned on ideological grounds. He came fourth in the finals (but later claimed he had been forced into the trick by his political masters).

To avoid a repeat of Dora's exploits, and to quash rumours that certain Russian shot-putters were not what they claimed, in the 1960s the International Olympic Committee began to test female athletes. At first the trial involved a simple 'nude parade' in front of doctors. Quite soon, a mere march-past was superseded by a deeper probe. It involved genetic and hormonal checks that were, to those who made them, unambiguous. The uncertain nature of development put an end to that idea. Several candidates (around one in four hundred in the 1996 Games) were deemed unworthy of a 'femininity certificate' and were forced to drop out. In public, most used a false claim of injury as an excuse.

The Polish-born runner Stanislawa Walasiewicz won the 1932 Olympic women's gold medal for the hundred metres and the 1936 silver in the same event. Fifty years later, as Stella Walsh, she was shot dead in a robbery in her new home of Cleveland, Ohio. At the postmortem examination she was discovered to have internal testes. In more recent years, Maria Martinez Patino, a Spanish hurdler, hoped to compete in the World University Games in Japan. At Femininity Control Head Office she too was found to have a Y chromosome. The unfortunate athlete was mocked in the press, lost her sports scholarship and was disqualified from the games.

After protests by her fellows, her career as a sportswoman was allowed to continue. Now the International Amateur Athletic Federation has called for the whole policy to be dropped, as it does little more than invade the privacy of people with unusual patterns of development. Not until the 2000 Games in Sydney did the committee agree (and then on a conditional basis).

The Olympians are, it seems, not convinced of the ambivalent nature of manhood. Millions of people, athletic or otherwise, masculine, feminine and of uncertain gender, prove them wrong. Sex is not an absolute, but depends on a web of hormones, of enzymes and of controls within the cell that can be torn in many ways. Failures lead to unexpected and often painful ends, but can also give a new insight into the equivocal nature of masculinity.

In 1838 a young girl was born in La Rochelle, on the west coast of France. Soon Herculine Barbin was enrolled in a convent. Her adolescence was ambiguous. 'I reached seventeen . . . At that age, when all a woman's graces unfold, I had neither that free and easy bearing nor the well-rounded limbs that reveal youth in full bloom . . . My features had a certain hardness that one could not help noticing. My upper lip and part of my cheeks were covered by a light down that increased as the days passed.'

Her memoirs, effusive as they may be, show the pain of abnormality. 'Alone! Forsaken by everyone! My place was not marked out in this world that shunned me, that had cursed me.' She began to share a bed with her friend Sara ('I would smooth the graceful curls of her naturally wavy hair, pressing my lips now upon her neck, now upon her beautiful naked breast!'). Herculine used her enlarged clitoris to such effect that a rather confused Sara suffered a pregnancy scare ('Sara made a disclosure to me that overwhelmed me! – If her fears were well founded, we were lost, both of us!').

Her diary describes the pains in her groin. A doctor was called. The answer was clear: she was a boy, and, as the physician noted, a potent one. 'At night, her dreams were sometimes accompanied by indefinable sensations. She felt wet, and in the morning found greyish and starchy stains upon her linen.' His mother had not lost a daughter but gained a son, albeit with undescended testicles and a half-formed penis; but the poor child was distraught.

He took up a new identity as 'Abel', left La Rochelle and Sara and moved to Paris. There he fell into despair and killed himself with the fumes of a charcoal stove. His diary records the 'incessant struggle of nature against reason that exhausts me more and more each day, and drags me with great strides towards the tomb . . . When that day comes, a few doctors will make a little stir around my corpse.' They did, and found the truth.

It has now been much extended. Herculine Barbin suffered from an inborn shortage of the enzyme that acts on testosterone to increase its potency (a problem shared a century later by Stanislawa Walasiewicz). The modified chemical is of crucial importance in the growth of the external genital organs. Any failure allows the female within to make an appearance (although, as in Herculine's – but not in Stanislawa's – case, the man waiting in the wings may emerge at adolescence when a surge of adult hormones pushes him, and his genitals, to centre stage).

Other girls with Y chromosomes – Maria Martinez Patino included – inherit an error not in the hormone or its special enzyme, but in its receiver. Some show no overt sign of abnormality at all and live as women, quite unaware of their ambiguous state.

The genes in control of androgen receptors are as liable to go wrong as are any others. The molecule which welcomes testosterone into the cell is coded for on the X chromosome. It can be altered in various ways, either by simple mutation or when extra pieces of DNA shoulder their way in. Some of the errors alter the pocket to which the hormone binds, while others harm the segment that attaches to a target gene. They can block, completely or in part, the road to manhood.

Such mistakes may cause no more than a slight loss of masculinity but can also lead to the birth of a child who is, at first sight, entirely female. The amount of male essence in such girls is that of a normal boy, but it never reaches its goal because the lock to the cellular door is damaged. As the material is transformed by aromatase into oestrogen, such people are often endowed with large breasts and good skin, with almost no body hair.

A second celebrated Parisienne, Wallis Simpson, the wife of the Duke of Windsor, may, rumour has it, have suffered from the problem. Her first name is of equivocal gender (perhaps a hint of a certain doubt about her natal identity) and she is – in spite of her three marriages – reputed to have claimed never to have been touched 'below the Mason–Dixon line' (fantasy plays a part here, for the Virgin Mary, Joan of Arc, Queen Elizabeth I and Marilyn Monroe have all been said to share the condition). For some people in her predicament, the first hint of difficulty arises when they find themselves unable to have children. For others, testes may burst through the body wall to show what has gone wrong.

Such faults are not rare. The term 'ambiguous genitalia' (much used by doctors) is in itself ambiguous, but one baby in three thousand or so has one or other of the several errors that can modify the reproductive organs. At birth or later, their state begins to reveal itself.

Once, such unfortunates were blamed for their condition. A certain Victorian lady who tried to claim a legacy as her genitals became more and more male was condemned to a lunatic asylum, her new appendage diagnosed as a delusion of masculinity. Later, surgery to restore what seemed normality – to boy or girl according to appearance and, often, at the whim of a doctor – became almost automatic. A standard text of two decades ago described the problem as a medical emergency to be resolved within a day, and parents were sometimes rushed into the decision to transform their child from boy to girl with almost no time for thought.

Such mindless meddling often led to tragedy for those whose mental and physical worlds did not overlap. Surgeons had their own preferences. For some, boy babies with a penis of less than a centimetre and a half were operated upon to remove any evidence of maleness. Others were stricter. Their patients needed an organ of two centimetres to save them from the knife – although they might have managed quite well in their original state. It is fatally easy to carve out a vagina, and many have been made, but the end result does not work well. Often the need for certainty led to disaster. More than half of those who undergo such operations return for further adjustment and some suffer lifelong trauma.

The new insight into the complex path from gene to sex, and from gender to identity, has forced medicine – and society – to become more tolerant. The notion of two distinct forms of humankind, each with an ordained way of life, has at last begun to disappear. Some children with ambiguous genitals are now allowed to develop, if they so desire, without interference. More and more, they are left to make up their own minds what they want to be (even if a few surgeons still feel that to leave them untreated is just a response to pressure from the adult transsexual lobby). In practice, a minority of those raised as girls decide to feminise their genitals when they become adolescent, but most brought up as boys are anxious for an operation to make their nature clear.

From Herculine Barbin on, such problems hint at the uncertainties of every embryo. They begin at conception. The early fetus is, whether or not primed by a Y chromosome, undecided as to its state. Sperm and egg are born outside its confines, and for a month or so everyone is neuter, but then the body is invaded by sex. Its several hundred germ cells (the precursors of sperm and egg) respond to a signal and – after a brief sojourn near the early heart – move to their future home. Dozens of genes help them on the way.

Once within their permanent abode, the germ cells, and the ambiguous tangle of tissues that in half of all babies becomes a penis, take the first – but still not irrevocable – step to the adult state. The gonad decides on its destiny and, most of the time, its owner follows. The *SRY* gene clicks into action about a month after fertilisation, when the embryo is about a centimetre long. Mice soon switch it off, but in men and pigs it broadcasts its message into adulthood. Why, nobody knows.

Soon a small block of tissue involved in water balance changes its career. That interesting lump is joined by the progenitors of the genital organs and in boys drifts away from the early kidney (which has a feminine touch for cells in contact with it). The testis and its hangers-on emerge from a sort of biological origami carried out with two tubes, the Müllerian and the Wolffian ducts. Their unerotic conduits are folded, cut and fused to make the adult genitalia: male from the organ of Wolff, and female from that of Müller.

Certain cells in the young testis make a protein that causes the Müllerian tube to commit suicide. At once the Wolffian canal sets off on its masculine tasks and, after a brief career as a mere guide of urine, shapes itself to form the pipelines, valves, reservoirs and treatment stations that help sperm on its way. Like most hormones, this Müllerian Inhibiting Substance has other jobs. It tells the testes when to move to their adult home and is needed even in girls, where it puts egg formation on hold until puberty.

As the inner surgery and sculpture go on, the external genitalia are born. The penis appears as an unimpressive smudge of tissue, a fold accompanied by a bulge. That humble structure is transformed into a phallus by the fusion of the folded section, while the bulge becomes a scrotum. The testes work best when cool, and stay within the body until just before birth, when they are drawn down to their home by a rigid ligament, the gubernaculum (named after the rudder on a Greek ship). Elephants do not bother with it and retain their organs within the body with no apparent problem, but boys in such a situation are sterile. In mice and other creatures the testes migrate in and out of the body with the season.

About a quarter of the way through fetal development comes a rush of testosterone. From then on, an unborn boy is marinated in a bath of masculinity. It allows him to race ahead of his sisters in the genital handicap, for ovaries are not defined until much later. For his first couple of months a boy makes, relative to his size, as much of the chemical as does an adult. Its first job is to push the Wolffian ducts further towards maturity and, with the help of emissaries from the brain, to refine the structure of his generative organs.

Cattle freemartins – the females who share a circulation with a male twin – showed how male hormones can, given the chance, work on the opposite sex. In mice, with their large litters, a female who develops next to a male alters in both brain and body compared to one with another female as a neighbour. The effects in ourselves are smaller, but girls with a boy twin might, some claim, have larger teeth and a more audacious attitude to life than those who share a prenatal home with a sister.

As growth goes on, man's molecules do more and more. Testosterone's modified and more active form causes the penis to lengthen by a dozen times in the six months from its first appearance and marks the newborn as an unequivocal member of his own sex. At birth, the proud parents check at once for its presence. Unseen by them, their son has a surge of hormones in his first few hours which persuades his organ to continue its career. From the age of three the testis, too, enters a decade-long phase of expansion. Its internal tubes increase from five hundred feet at birth to twelve hundred at the age of ten, and the number of cells ready to become sperm shoots up to match.

Then comes puberty, with the familiar joys of acne, gruffness and desire. It seems an abrupt stage in the seven steps to manhood, but its agenda is set well before any obvious signs appear. Gonadotrophin-releasing hormone is the key. It surges in great pulses from the brain, and pushes up the level of testosterone. Young men need in addition a dash of femininity in order to reach their allotted state, as those with mutations in the gene that makes their oestrogen fail to mature.

The choristers of King's College, Cambridge, are famed for their performances of Bach. Their ethereal voices last, on average, until the age of thirteen. When their forerunners at Leipzig first sang the music two centuries earlier, the boy sopranos kept their trebles for four more years. Since then there has been a dramatic decrease in the age of maturity. *Homo sapiens* is still slow to become adult (even blue whales do it quicker), but boyhood has become shorter and shorter (and several twelve-year-olds have become fathers). Its abbreviated state reveals a connection between the web of maleness and the chemical networks in control of the rest of the body.

Restaurants are well aware of the erotic potential of a good meal. Food sets off, within a few minutes, a squirt of gonadotrophin-releasing hormone from the brain and, to follow, a pulse of luteinising hormone from the pituitary. Starvation and intense exercise each turn both chemicals off (which is why woman athletes tend to lose their sexual cycles). The link between dinner table and double bed is made through a protein called leptin, which was discovered when a mouse that ate with much enthusiasm and became bizarrely overweight appeared in a laboratory strain. It bore a new mutation called *obese*, which damages a messenger made in fat-storage cells whose normal job is to sense how much spare energy is available and to tell the body when to divert its efforts to secondary interests such as puberty.

Today's choristers eat better and tip the scales sooner than did those of eighteenth-century Germany. Since Bach's day, children have become adult with fewer years, but at about the same number of pounds weight. The end of boyhood is set, in part at least, not by time but by gravity. Fifty-five kilograms of solid flesh are needed before adulthood can rear even a tentative head. As soon as the hormonal weighing machine tells the testes when the goal has been reached, leptin helps to switch the body's efforts into maturation rather than growth alone. The rare children who bear mutations in the leptin gene are – like *obese* mice – very fat and do not become mature. In some animals, maturity needs more than food (long days or the presence of a mate may be called for); but for ourselves, avoirdupois is the key.

Several other changes happen at that time. Boys shoot up in height, and a surge of androgens causes body hair to appear. Within the skull the amount of white matter goes up (more in boys than in girls) while the dose of grey matter – the seat of intelligence – rises but then falls. It reaches a peak among boys at thirteen or so, a year later than in their sisters. At last, a treble's voice breaks, and it is time to leave the choir.

Once through the pubertal shift, a youth embarks on his adult career, the fifth era of life, with its familiar and repetitive complications. For a long age, testosterone is at centre stage. Its levels vary from one individual to the next by ten times, and for some (but not all) fluctuates over the course of the day. In the Congo, it drops during the annual starvation period, and the springtime peak in condom usage in Britain (as manifest in the records of sewage works) hints at an annual cycle in the developed world as well. Then comes an unwelcome change.

Various eccentrics have kept erotic diaries and, unreliable as they may be (some claim a hundred thousand ejaculations in a lifetime, most of them self-induced), all show a common pattern. Activity stays high until the mid-thirties, and then declines. Androgens show the same shift. Women also have hormonal changes, but theirs are abrupt, with a rapid shutdown at the menopause. Do aged husbands share their wives' experience?

Some physicians identify a 'male climacteric' – from *klimacter*, the rung of a ladder – a definite step downwards on the descent to old age. Hormones do change as the years pass by, and at the age of seventy testosterone is down by half from its peak (it reaches a low point in the late evening, just when it might be called for). The vital fluids also suffer. The concentration of sperm stays about the same, but the amount of ejaculate made decreases. It contains more abnormal cells, and even the normal ones find it harder to swim. As a result, the fertility of fifty-year-olds drops by a third compared to that of males of thirty. A ninety-four-year-old is known to have fathered a child, but long before then most of his fellows have lost interest in copulation and have hot flushes and genital flaccidity to match.

The sad lack of fit between biology's insistent urges and the disorders of age goes back two thousand years. Once, no help was needed, as Adam lived to nine hundred and thirty after his expulsion from Eden (and his grandson Enos fathered a child when he was ninety), but then things went downhill. King David, when stricken in years, turned to the latest therapy. His servants had an idea: 'Let there be sought for my lord the king a young virgin . . . and let her lie in thy bosom, that my lord the king may get heat.' The young lady was Abishag the Shunamite and although – as the Good Book puts it – she was very fair, and cherished the king, and minis- tered to him, the king knew her not. In spite of its failure to cure David's impotence (which led to the loss of his polit- ical power) the practice was prescribed by physicians as late as the eighteenth century. The idea stayed popular, among old men at least, for much longer.

Shunamites are hard to find today, but many elderly males still try to reverse decay. The Chinese once used extract of testis to treat those short of 'masculine activity' (a measurable effect calls for a dose of five pounds of bull testicle a day). Old men are no doubt honest in their complaints but most, alas, come from simple age rather than in response to some sudden change in chemistry. Much as testosterone may help boys in sexual difficulty, age is beyond its reach. It does help some patients to lose weight, but makes no difference to the mood of the elderly, or to their physical strength. Even in ninety-year-olds, its levels almost never fall low enough to interfere with the reproductive machine.

It can be dangerous to meddle with the androgens. Males have a large gland near the base of the penis whose job was once unknown: as the politician Clemenceau put it, there are only two useless organs, the French presidency and the prostate gland. Whatever its function, the organ is a cause of distress to some elderly men. It often becomes swollen and interferes with the ability to urinate. Some unfortunates suffer a more serious problem. Prostate cancer, if not caught in time, can kill. In the 1930s it was treated, with typical medical hubris, with a drug called Hombreol – in fact testosterone itself (also recommended for schizophrenia). As the tumours are in fact encouraged by that substance, the therapy was misguided, and to take it in the last age in the vague hope of an increase in good cheer is too risky, whatever the lure of youth restored.

The essence of maleness is more useful to aged women (who do not even own a prostate) than to their partners. Testosterone peaks in a woman's twenties, and by her forties is already down by half. After the menopause, or the surgical loss of the ovaries, some lose all erotic interest. A dash of the tincture of manhood added to their hormone replacement therapy brightens up their lives, with more intercourse and more orgasms. Perhaps, some say, it could be offered to all females of a certain age in doses small enough to cheer them up but to avoid the threat of suffering, in some senses, a change of identity.

Men themselves evolve towards their wives as they age. The enzyme that transforms their prime hormone into oestrogen increases in activity with the years – which is why the ancient seem almost neuter, with a voice that turns again towards childish treble as the woman within at last makes her presence felt.

And what of the most ignominious fate of all, shared by the old of both sexes? When the thinning Napoleon met the almost hairless Czar Alexander of Russia to discuss the future of Europe, they talked instead of baldness cures. For emperors and others, the best treatment was once to do nothing, or to turn to wigs. Those anxious to avoid artifice (or, indeed, hair) could use bat milk, have their scalp licked by a cow or invest in 'Hair in a Can' (spray-on chopped wool). The Egyptian nostrum of a blend of the fat of a lion, a hippopotamus, a crocodile, a cat and a serpent is just as effective. Surgery is brutal but better. A flap of flesh is peeled from the side of the head and shifted to the top, or a thousand holes are punched in the scalp and small groups of hair transferred (the 'living wig' of gold wires woven into the skin with artificial curls attached is not recommended).

Soon operations may be out of date because the drugs have begun to work. Minoxidil ('Stronger than Heredity!') was invented to help people with heart disease. For some, it had the unexpected side-effect of holding back the tide of scalp (quite how, nobody knows). Another drug, Finasteride, developed for prostate problems inhibits the enzyme able to convert testosterone into its more potent form. In lower doses it, too, assists those who lose their hair. Such therapies are expensive, at several hundred dollars a year for life; each works in less than half of those who try them; and neither is of use to the already bald. A new and more potent version of the prostate drug is said to work miracles on the top of the head (and also increases the risk of impotence) but has not yet been passed as safe for use.

Julius Caesar, Napoleon, Lenin and their depilated fellows can console themselves with the thought that their condition is a statement not of premature decay, but of the subtlety and power of maleness. They should not mourn their pates but should instead rejoice in their good fortune at having achieved, against the odds, its crowning glory.

CHAPTER 4

HYDRAULICS FOR BOYS

Modern technology proves what passes across an unborn baby's mind. An ultrasound probe shows that most boys, a month or so before birth, have an erection for about an hour a day. Twenty years on and the habit is well established, for nearly all young men are in a state of physical excitement for three hours out of every twenty-four. That discovery also needed a mechanical device: a snap-gauge in which a plastic thread breaks when under tension, or a RigiScan, a hydraulic band which tightens every thirty seconds to tell the operator how engorged his subject's organ might be.

The equipment is needed because most erections happen while their proprietor is asleep. Woody Allen's brain may be his second favourite organ, but the majority of his, or anyone else's, penile displays are outside its conscious control. They hint at the subtlety of the masculine machine. Its most honourable member is controlled from afar down a precarious line of command and breakdowns are not rare. More than half of males over forty have some sexual difficulty and, twenty years on, one in six is quite unable to maintain an erection.

Such problems are not new. Hippocrates wrote of the Scythian nomads of the steppes that they became impotent because they were exposed to the constant jolting of their horses. Injury puts a stop to many male ambitions (cyclists in particular suffer as blood flow to the crucial organ is reduced by two-thirds as soon as they sit on the saddle), as do old age, illness and despair. Diabolic possession was thought to be to blame, and pictures of witches often showed sausages strung limply over a stick. Exorcism was then the only solution.

It was succeeded by a variety of quasi-scientific treatments. In 1780, James Graham opened his Temple of Health and Hymen in the Adelphi in London, close to where the Savoy Hotel now stands. Its board of management included Lady Hamilton, mistress of Lord Nelson, and it offered a Celestial Bed, powered by static electricity and aimed at the impotent. 'Twelve feet long by nine feet wide, supported by forty pillars of brilliant glass of the most exquisite workmanship, in richly variegated colours', it was under the control of a certain Hebe Vestina, the Rosy Goddess of Youth and of Health, who sold in addition her Nervous Aetherial Balsam.

Purification, galvanism and nervous ointments have been superseded by a huge industry that turns on man's desire to stay upright. Some of its products are no more effective than was the casting out of devils, but science can now persuade many reluctant organs towards the vertical.

Thomas Browne in his *Religio Medici* of 1643 called the 'trivial and vulgar way of union . . . the foolishest act a wise man commits in all his life'. That might be so, but a failure to be foolish can devastate the lives of even the wisest. Problems both physical and metaphysical may subdue desire and reduce performance. They come from disease, from damage or from inner demons. After years of charlatanry, medicine has begun to come to terms with the sexual ills of both body and brain.

Love starts with chemistry but it ends in physics; as an exercise in pumps, valves and fluids. Like a hydraulic braking system, the mechanism is in essence simple, but in practice complex. When it comes to coition, men have a harder time than most, for they depend on hydraulics alone while almost all other mammals have a bone to shore up their most strenuous moments. It allows the penis to be inserted before it is quite erect, and may also enable a male to remain within his mate for as long as she will permit (weasels manage two hours at a time) rather than being limited by his own ability to keep up the pressure. The support bone of the blue whale is a yard long, but man himself needs only plumbing. Sometimes, the fluid mechanics fail and flaccidity follows.

The organ spawns obsessions. Cupboard fifty-five in the British Museum is the Secretum, the legacy of George Witt, Mayor of Bedford and collector of phalluses in stone, wax, amber and bronze. In 1865 he donated to the museum his life's work: a collection of hundreds of objects 'symbolic of the early worship of humanity' (which, he thought, had once been entirely based on fertility rites). Some stand alone, while others have bodies attached, or wings, or small engraved birds. One of the most remarkable belongs to Jesus Christ, shown erect as its owner is crucified. The collection was added to throughout the nineteenth century by, among others, Sir William Hamilton, the Nelsonian cuckold. Then the cupboard was closed except to those 'of taste and discrimination'. Now its contents are, like most of man's secrets, on open view.

Some of Witt's relics are a yard long, but such things are exceptional. The *Kama Sutra* recognises three sizes: hares, bulls and horses. A survey by an American condom manufacturer came up with an average of 5.877 inches (a confident figure, but smaller than previous reports). More than half the erect organs were between five-and-a-half and six-and-a-quarter inches, and nine out of ten claimed more than four-and-a-half inches. Surgeons themselves regard any penis over three inches' erect as normal. A study in a Japanese brothel gave a mean of five-and-a-half inches, and – given the technical errors inevitable in such places – oriental and occidental members are probably much the same size.

For artistic reasons and to emphasise a subject's moral status, sculptors often reduce them. The earliest Greek statues, the *kouroi*, are accurate in most proportions, but have tiny penises. As their testes are of normal size the images look odd to modern eyes. Michelangelo's *David* – defiantly uncircumcised, albeit a Jew – has an appendage no longer than his thumb (although he does have large hands). Today's artists are as likely to exaggerate as to diminish the male organ (or, as in the photographs of Robert Mapplethorpe, to do away with the owner altogether to emphasise its bulk). In a statement of modern aesthetics, the copy of Michelangelo's great work at Caesars Palace in Las Vegas has been fitted with an augmented part to satisfy an audience used to reality on the Internet.

Some of those at the tables have, no doubt, undergone phalloplasty – penis-enhancing operations – most of which involve a simple injection of fat to thicken the structure. A Danish device uses a plastic rack and pinion to stretch it, in much the same way as certain African peoples extend the necks of their girls with a series of metal rings. It has some success. Apart from a complete and painful reconstruction of the kind offered to those with severe physical problems, most surgical attempts to lengthen the penis are misguided, as they do no more than cut a ligament at its base and cause it to droop further from the body cavity.

Whatever their size, all such structures contain three rods of tissue, surrounded by a series of wrappings and supported by a web of elastic fibres. Two are formed from a net of flexible reservoirs, held in a mesh of smooth muscle (a tissue also found in the intestine and elsewhere and not under the conscious control of the brain). The third rod lies between them. Within it runs the urethra, a tube for urine and semen. The tip expands to form the glans or 'acorn', which is in turn covered by a sheet of flesh called the foreskin or prepuce.

Blood enters the expectant member through a system of arteries which break into hundreds of corkscrew-like vessels. These internal reservoirs are the engines of erection. They drain through a system of veins, some deep within and others on the surface. The process is controlled by several sets of nerves, most of which work in pairs, some proposing a certain action while others oppose it.

Man's rise and fall depends on the ebb and flow of blood, which is in turn controlled by the tension of the muscular valves around the arteries of the erectile rods. For most of the time they are closed, but when they relax, the vessels open. Blood rushes in at fifty times its normal rate and fills the reservoirs (whose walls have become slack in readiness) with predictable results. As the erection grows it feeds on itself. Inflation compresses the inner veins and helps to block the drainage channels. This keeps the vital fluid where it is needed and maintains a solid state. To prime the penile pump needs around a teaspoonful of liquid each five seconds but once the perpendicular has been achieved the organ can manage on half as much.

The penis is not alone in its response. The testes swell and move upwards, the nipples become erect, the heart beats faster and pushes up blood pressure, and the buttocks of some fortunate men turn red. As soon as a discouraging signal is sent from the brain, the crucial valves close. At once the arterial supply fails, blood rushes out and the virile member droops.

Man's most intimate machinery, like an advanced braking system, relies on a central command post and a battery of local controls. The control chain has substations in the hypothalamus at the base of the brain and in the spinal cord itself. Its mechanism involves nerves and their chemical messengers, a variety of hormones and a set of changes within the cell. An erection is a subtle thing, with several independent lines of command, not all of which are associated with sex. Erotic ideas may spark it off, but most of the nocturnal events have nothing to do with libidinous dreams. Testosterone increases their incidence, but has no effect on those provoked by desire. There is, it seems, more than one path from brain to phallus.

Some of the byways are unexpected. In rats, a certain group of brain cells is half as large again in males as in their opposite numbers. It takes messages from the nose – and scent means a lot to a rat – and passes them on to the brain structures associated with erection. In men, too, that section increases its activity as excitement begins, as a hint of a tie between nose and genitals. Mice can tell how related another animal might be from its smell. Females shun the advances of their close kin – and they are wise to do so, for the sons of such matings are sexual failures when released into the wild – and scent tells them which males to avoid.

Most of our own scent genes have lost their function, but the price of perfume (based, as many are, on the mating odours of animals) suggests that a few still work. Sordid experiments with soiled T-shirts hint at an ability to separate our fellows in this way (and at a preference by women for partners who smell rather, but not too much, like their own fathers). Such choices might play a part in the choice of a mate, as some boys born without a sense of smell are not attracted to the opposite sex.

The state of the genital organ depends on a balance between several sets of nerves. The battle for inflation has two main players. The parasympathetic nervous system, as it is called, summons up the blood with chemical messengers sent across the gaps between nerve and muscle which cause the arterial valves to relax. They are countered by the sympathetic system, which plays a rival and unmanly part.

Quite which messenger sparks off the excitement at the business end of the operation is not altogether clear. Most parasympathetics use a substance called acetylcholine to cross the gap between the point of arrival of an electric message and the muscle to which it is directed. Acetylcholine is, without doubt, involved in the day-to-day economy of the penis. However, drugs such as atropine (a poison found in deadly nightshade) that block its action have but a small effect on erectile powers, which hints that another chemical may bear the crucial news from nerve endings to muscles.

Once the tumescence has done its job (and sometimes before), central control sends out a call for moderation. The sympathetic nerves then come into play. They pump out nor-adrenalin (a hormone involved in the fight or flight response brought on by a sudden alarm). This causes smooth muscles to tighten and the arterial valves to close. At once, the gates swing shut and the apparatus returns to its trivial self, where it stays until the balance of control shifts again.

Many other nerves serve the masculine member. Some pass the frictional message from its sense receptors back to the brain. Others tell the ejaculatory muscles to do their job. In addition, the penis is as well supplied with fibres able to perceive pain as is any other organ. The nerves call on various intermediaries. Prostaglandins act rather like hormones as they push a target cell into action, but work close to home rather than at a distance and – unlike most of those chemicals – break down almost at once. They, too, relax smooth muscle. Other chemicals, including a mysterious group of small proteins, assist in its rise and fall.

In 1998 the erectile organ attracted a Nobel Prize. The award was established by a wealthy Swede to assuage his guilt at the havoc wrought by his company's products. Alfred Nobel's great invention, dynamite, was a mixture of silica and a high explosive called nitroglycerin. In his old age, Nobel himself suffered from angina, a constriction of the blood vessels of the heart. A year before his death in 1896 he commented on the 'irony of fate that I have been prescribed nitroglycerin, to be taken internally. They call it trinitrin, so as not to scare the chemist and the public.' He refused to take his medicine, and paid the price. Quite why a high explosive should help a heart patient, nobody knew.

His own employees gave a hint of the truth. Those on the production line often complained of severe headaches. In 1986 the key to Nobel's heart and to his workers' heads was found. A simple gas, nitric oxide (until then notable only as a component of acid rain and in part to blame for the destruction of the ozone layer) emerged, to universal amazement, as an essential messenger within the body. The gas passes through cells and breaks down as soon as it has been used. Simple as it may be, it is a central player in every man's economy.

Its powers were first identified in blood vessels, where nitric oxide causes walls to relax and flow to increase (which explains both the headaches and the ability of nitroglycerin – a source of the gas – to help a heart starved of oxygen). The chemical is also the force behind amyl nitrite (a substance related to the explosive and sold on the street as 'poppers') which makes arteries slacken and gives an instant rush of blood much appreciated by drug users.

The vital gas is made by a special enzyme which breaks down an amino acid called arginine. The reaction is simple enough, but the gene for the enzyme involved is the most complex ever found, with a variety of on–off switches and an intricate system of internal controls that summon up certain parts of its structure in some tissues and others elsewhere.

Nitric oxide does many jobs in many places: in nerves, in blood vessels and in the immune system. Rats infected with bacteria use it to kill the invaders. In the brain, the chemical is everywhere, and is implicated in memory, sleep, pain and depression. Male mice in which the pathway has been damaged by mutation are more aggressive (although, oddly enough, females are less so). Among its other tasks, the gas is an on–off switch for the lights of fireflies. Nitric oxide is now much used in medicine, where it helps babies with lung problems and is useful in diabetes, high blood pressure, loss of memory, and even sunburn.

The famous gas has its highest profile in the most masculine of organs. There the substance acts as a final link in the chain of command. A signal from the nerves sparks off a puff of nitric oxide, which in turn triggers off the cascade of changes that cause the valves to let in the hydraulic fluid. As the rampant penis grows, the arteries themselves continue to pump out nitric oxide and to maintain its stance until retreat is called for. Then the sympathetic nerves come into play. The nitric oxide is broken down by a second enzyme, the arteries and valves close down, and the excitement dies down.

A successful love life needs more than erection alone. Sperm has a long way to go before it leaves its owner. A thick muscular tube – the vas deferens – connects each testis to the base of the urethra. When the time comes, this contracts to push the fluid onwards. At the base of the prostate gland it meets a reservoir called the seminal vesicle. From there the tube continues, to form an ejaculatory duct. At the right moment a mass of muscle, which extends an inch or more around the base of the male member, contracts and expels the seminal fluid. The process is controlled by the somatic nerves – the system associated with conscious movements – which is why men tend to enjoy it. Ejaculation has three phases: fluid is squeezed into the base of the urethra; the neck of the bladder closes to avoid a last-minute wrong turn; and at last the sperm is ejected. From then on, the liquid is on its own.

Hydraulic systems, be they in giant excavators or modest penises, work well most of the time, but in both devices a tiny fault can bring the entire machine to a halt. Their failure has long mystified and mortified mankind. As Montaigne said in his *Essays*, 'The indocile liberty of this member is very remarkable, so importunately unruly in its timidity and impatience when we do not require it and so unseasonably disobedient when we stand most in need of it: so imperiously contesting in authority with the will, and with so much haughty obstinacy denying all solicitation, both of hand and mind.'

Impotence (or erectile dysfunction, as doctors now call it: the earlier term has been purged to help men to discuss their problem) has many causes. It was once thought to come from masturbation, a habit to be avoided at all cost (even if Diogenes indulged in public to avoid over-full testes; as the philosopher said, 'I wish to heaven that I could satisfy my hunger by rubbing my stomach'). A pamphlet of 1715 entitled *Onania, or the Heinous Sin of Self-Pollution, and its frightful consequences in both sexes considered, with spiritual and physical advice to those who have already injured themselves with this abominable practice* claimed that 'those who are guilty of it work for the destruction of their own species and strike a blow, in a certain way, against the very Creation'. A century later, Benjamin Rush, father of American psychiatry, wrote that 'Masturbation produces seminal weakness, impotence, dysury, tabes dorsalis, pulmonary consumption, dyspepsia, dimness of sight, vertigo, hypochondriasis, managlia, fatuity, and death.' In 1994, in contrast, the nation's Surgeon General saw the habit as part of human sexuality, which might even be taught in schools. She was at once dismissed by President Clinton.

The Surgeon General was right. Erectile failure has causes more profound than self-abuse. A third of men under sixty suffer to some degree, and the figure rises with the years. A mere one case in ten comes to medical attention, but the ageing population of the Western world will cause the numbers to double in the next two decades, to three hundred million worldwide. Its causes are various, some avoidable and some, alas, not.

Men with high blood pressure are often unable to summon up enough hydraulic fluid to pump up their organ. Smokers suffer from the problem at several times the normal rate (which was the basis of a successful anti-tobacco advertisement, in which three men turn to look at a girl and their cigarettes go limp). Alcoholics and keen consumers of marijuana face the same risk. A stroke, or multiple sclerosis, can also put an end to erotic activity. Half of all diabetics suffer as their muscles lose the ability to respond to a nerve signal. Even worse, a quarter of those on prescription drugs must face the issue because the medicine interferes with the reproductive apparatus. Class, too, interferes with the libido, for men in manual jobs suffer more from the ailment than do managers, even when other factors are taken into account – indeed, to be poor unmans a male almost as much as do alcohol and tobacco.

The genital pipework, like that of a motor vehicle, wears out with the years. Old age reduces the elasticity of arteries, skin, muscles and – as a result – penises. As its tissues become less pliant the sluices can no longer close, and failure follows. The passing years tax other parts of the male machinery as well. Some men notice a reduction in the force of their urine stream and, by the age of sixty, half have the problem. The prostate has made its presence felt. In many of those over forty the organ is enlarged and some with the problem go on to develop cancer. The first sign is often a disastrous inability to urinate, followed by an operation. Just one in eight males knows where his prostate is (and more women are able to locate the crucial structure than can those who actually own one) and, although surgery is more sophisticated than it was, many lose their powers after it has been operated upon.

Other unfortunates suffer from Peyronie's disease – a fibrous mass that may become hard and chalky. It can cause the penis to bend when erect, or to refuse to extend itself at all. The cause of the condition is not known. Sometimes the lump disappears without help, but if it does not, steroid creams, ultrasound to break up the obstruction or surgery may be called for.

However they arise, the extent of a patient's problems can be assessed with the fifteen questions of the International Index of Erectile Function, or Europe's terser Brief Sexual Function Inventory. A low score means that no longer does the RigiScan bear its message of excitement and, quite often, its despondent subjects turn to medicine for help.

Some of the remedies on offer are crude while others depend on the latest advances of science. Sometimes, a plumber can sort matters out, with surgery to clear a blocked artery or to tie off a vein to reduce leakage (although the blood has the inconvenient habit of finding its way around such artificial barriers). Implants, of disagreeably unbending nature, were first used in the 1940s. Thirty years on, with the development of flexible silicone, medicine turned to semi-rigid rods and to inflatable replacements for erectile tissue. The rods are made from a supple core of silver wire covered with soft plastic, while the expandable versions depend on a reservoir of salty water and an external pump. The most recent models come in several sections to give the flaccid member a more natural appearance, and are served by a pump in the scrotum and a reservoir hidden in the abdomen. In rats, implants have been grown from the body's own cells on a template of degradable material, and this may soon be available for those who prefer flesh to plastic.

Most users of such devices find them helpful, and the failure rate is low. Even so, inserts can cause problems. They can move to where they are not wanted – even out of the member itself. As the implant wanders, it may interfere with the ability to urinate. An inserted shaft can be painful to the patients who receive it, and the permanent semi-erection enjoyed by those so adorned is less than elegant. Some patients also suffer what surgeons call 'Concorde Deformity', in which the tip droops in an ungraceful manner because the glans is not engorged although the rest of the penis appears to be.

For those not keen on the knife, a recalcitrant organ can be inflated with the help of a motorised or manual vacuum device. A contrivance to suck men to rigidity, a glass tube called the Peniscope, was patented in 1917 by the champion bodybuilder and health-food magnate Bernarr Macfadden (born Bernard, but his new name sounded more like the roar of a lion). He later founded a religion and failed to obtain the Republican nomination for US President.

Macfadden's device has stayed more or less unchanged since then (even if modern versions have a safety valve to prevent damage caused by over-fervent use of the pump). As the pressure drops, blood rushes in and tumescence follows. The hydraulic fluid must be kept in place with a tight ring around the base of the shaft, as otherwise it will leak away. The method works but can be clumsy – and the penis must be allowed to return to normal within half an hour as it may become starved of oxygen. In addition, some partners object to the insertion of a cold member, which loses heat because no blood is circulating.

The *Kama Sutra* promises a month-long 'enlargement' to those who try an infusion of ground cherry, sweet potato, nightshade, fresh buffalo butter, teak tree leaves and heliotrope. Fifteen centuries on, the publisher of *Onania* made similar claims for his Strengthening Tincture at ten shillings and Prolifick Powder at twelve.

Such things were succeeded over the years by a variety of chemical nostrums which had about the same scientific basis as those of ancient India. The medical principle of *similia similibus* treats a faulty organ with a functional version of itself. Its logic led the Romans to use donkey penises as an aphrodisiac, and the Chinese still rely on potions such as Essence of Tyrant and Big Hero Pill, each based on the same principle. Now, an era of drugs that actually work has arrived. Today's Strengthening Tinctures put heart into husbands, and vast wealth into the pockets of pharmaceutical companies.

People in the most urgent need are happy to inject their tonic straight into where it is needed. Papaverine, a substance obtained from poppies, is used in this way, as is nitroglycerin itself. Artificial prostaglandins, mimics of the natural relaxers of smooth muscle, do the job even better, and with commendable speed. Nine out of ten who use the needle are satisfied at first but half of them abandon the syringe within three years as their skin becomes more and more perforated. The latest technology involves a drug-filled pouch placed in the scrotum, with a vessel leading to the penis. A subtle squeeze at the right moment, and it springs into life.

Those who use such devices face the risk of priapism, a stubborn erection (cocaine can have the same effect). Sufferers must turn at once to the ice bucket, or inject an antidote to shrink the inflamed part. At worst, a doctor must be called to drain blood from the engorged tissues before gangrene sets in.

To avoid injections some patients insert a capsule filled with the drug into the penile tube instead. The Medicated Urethral System for Erection – a tenth MUSE to add to those of poetry, music, dance and the rest – does not work well and can be painful for the other member of the team when the drug leaks out. A prostaglandin cream to be rubbed onto the appropriate spot, as recommended by Diogenes, is also on the market, as is a container filled with the chemical, to be thrust into the urethra. Testosterone itself helps just a minority of the impotent. For them it is now available in a drug-filled patch to be glued to the scrotum to get the magic bullet as close as possible to its target.

The future of the phallus was transformed in 1998 by Viagra, the famous blue pill. This, the most profitable drug ever sold, began as a heart medicine but had little success. Then, in a trial in Merthyr Tydfil, there was an unexpected side-effect. The local Welshmen complained of involuntary and lengthy erections. At once, the pharmaceutical company Pfizer became interested in a new use for their product. Their success was such that they won an award for industry from the Queen.

Now Viagra is prescribed three hundred thousand times a year in Britain – and the United States, with a population five times our own, needs fifteen times as many doses. The chemical brings in more than a thousand million dollars every twelve months. So concerned was the National Health Service about the cost that the Minister of Health at the time referred to it as a recreational drug, whose availability should be restricted. Most of those who use it have to meet the bill from their own pockets, but at around ten pounds a pill on the Internet few are dissuaded on grounds of cost.

Sildenafil citrate, to use Viagra's technical name, inhibits an enzyme in the pathway that breaks down nitric oxide. The chain's normal job is to tighten smooth muscle, to cut down the organ's blood supply and to keep it flaccid. Sometimes, the enzyme can, if so minded, continue to work even when it is not needed. The penis then stays drained even in the face of instructions to the contrary. Viagra inactivates the guilty protein and allows the male member to respond as it should to central command.

Its target comes in various forms, each present in a different tissue and each with its own biochemical profile. Viagra works hundreds of times harder on the version of the enzyme found in the penis than on the forms made in heart and muscles (which is just as well, given the demands made on those systems when the pill does its job). Its great strength is its ability to act on the normal machinery of genital excitement. In effect, it prevents a premature flick of the 'off' switch which puts paid to so many masculine intentions. The drug, in the amounts prescribed, seems harmless. Taken an hour before intercourse it helps two-thirds of those who try it. Viagra has some side-effects – headache, flushed cheeks and a stuffed-up nose – but few complain. Some of its devotees begin to experience their sex lives in technicolour. For them, life goes blue during copulation as the drug binds to a version of the enzyme found in the eyes as well as that made lower down.

The new panacea has also been of indirect help to the males of other animals (although an attempt to use it to improve the breeding of captive pandas did not work and their keepers had to turn to instructional films), for it has cut the demand for Big Hero Pills and their relatives. A seal penis now costs a fifth of what it did and there is hope even for the more potent members of the tiger and the wolf.

Nine out of ten human users of the magic medicament stay faithful to it but some complain that after a couple of years (and the drug has not been around for much longer than that) it begins, like so many erotic aids, to lose its effectiveness.

The triumph of Viagra could be the dawn of an era of narcotically enhanced copulation. Several candidates have already presented themselves. The Eli Lilly Company's Cialis works on the nitric oxide pathway, and does so faster than Viagra, with an effect that may last a day. Some drugs go straight for the brain. Ixense, developed by a Japanese firm, was approved in 2001. It stimulates an erectile centre, and a pill under the tongue (or a nasal spray) works its magic in a mere twenty minutes. The occasional side-effects of nausea and fainting do not deter most of those who try it (to use it while drunk is not recommended because of the risk of a drastic drop in blood pressure). Ixense has been approved in Europe, but in the United States the Food and Drug Administration is still dubious about its safety. Patients with Parkinson's disease – a disorder of movement – often lose their desires. Levodopa, a drug that helps their muscle problems, also makes a difference to their sex lives and it might be useful to others.

Many treatments turn more on faith than on pharmacology. Yohimbine (obtained from the bark of an African tree and sold as herbal Viagra) acts to increase heart rate and blood pressure. It makes those who take it irritable, and even if it may help feed their vital part with what it needs, its effect is marginal at best, and may even be toxic. Fake products, too, play a part, and herbal remedies with names such as Väegra and Alprostaglandin have been forced off the market after bringing in around twenty million dollars a year. Sextasy – an illegal mix of Viagra, Ecstasy and amyl nitrate – works, but is not recommended.

Viagra can help in the first act of the erotic ritual, but things often go wrong further down the line. Some men can become erect, but then fail to ejaculate. Certain tranquillisers inhibit the process, while patients with spinal cord damage or with cuts in the nerves as the result of surgery can also suffer. Operations on the prostate, too, may harm the neck of the bladder and divert sperm from its proper path.

Damaged spinal nerves can be bypassed with a vibrator applied to the penis, or with an electro-ejaculator, a device first used by farmers, but now an important aid for the injured. The probe is placed in the rectum. Twenty jolts of twenty volts are enough do the job. The current stimulates the nerves of the urethra until semen dribbles out, ready for collection. For those with bladder problems, errant sperm can even be rescued from the urine.

Sometimes the generative fluid is frustrated because its path to the outside world is blocked. Some boys are born without a vas deferens, which at once prevents the passage of semen. Almost all those in this predicament carry a single copy of the cystic fibrosis gene which, when present in double dose, leads to severe inborn illness. As one European in twenty-five is a carrier, the gene is an important cause of reproductive breakdown. Others suffer because their ducts are blocked by the after-effects of infection or injury. Their problems can be helped by surgery.

In the world of sex, over-enthusiasm can be as much a problem as is failure. A squeeze at the base of the penis at the right moment may delay matters, as can a rub with an anaesthetic cream to slow down the whole process. Premature ejaculation comes, most of the time, from some emotional difficulty. An anxious male may also generate too much activity in his sympathetic nerves, which in turn tightens the smooth muscles, reduces blood flow to the crucial parts and prevents tumescence. The problem lies in the mind, but he may be cured with the same drugs as those used on patients with physical illness.

About a third of all cases of impotence come from bereavement, despair and shock, or from undue familiarity with a long-term partner. Nine depressed men out of ten have complete erectile deficiency, and to make matters worse, some antidepressants themselves interfere with male plumbing (although the wonder-drug Prozac helps it on its way). A whole industry has grown up around such people. Thirteen sessions of the Vienna Psychoanalytic Society were concerned with impotence. Some patients find cures of their own: Richard Krafft-Ebing, in his *Psychopathia Sexualis* of 1886 (a book at once perverse and tedious), described a case in which a husband could not become excited by his new wife unless she wore a gigantic wig. He had to choose the colour and style, and the device worked for a mere three weeks before it had to be replaced by a new model. As Krafft-Ebing reports, 'the result of this marriage was, after five years, two children, and a collection of seventy-two wigs'.

Quite how wigs, or psychotherapy, work is unknown to science, but such approaches can, without doubt, sometimes be effective. A sympathetic discussion with both partners is often useful, and if even that fails a surrogate may be called in.

In 2001 Pfizer sold more than twenty million Viagra tablets, as proof that the market is happy to invest large sums to avoid sterility. In the same year, the world used hundreds of times as many condoms, most of them bought by men anxious to do exactly the opposite (although with fifty million abortions a year worldwide, they do not always succeed).

Males have long been anxious to detach their erogenous means from its biological ends with whatever methods are allowed. The Church is just as insistent that the link be retained. Its view of contraception turns on a narrow interpretation of a famous text. Judah's first-born has died and he turns to his younger son, Onan, with an order to 'Go unto thy brother's wife, and marry her, and raise up seed to thy brother.' When the young man 'went in unto his brother's wife, he spilled it on the ground lest that he should give seed to his brother'. The Lord was less than pleased, and slew him.

Such is the basis of Catholic opposition to birth control. The sin, on other readings, has more to do with Onan's defiance of his father (a capital crime in divine eyes) than with his erotic habits. As Judah himself had a child by his daughter-in-law, conceived while she was disguised as a prostitute, the fate of his son seems unfair, but on such matters does the destiny of millions rest. The Pope recommends self-denial instead. Intercourse was once forbidden on Fridays (the day of Christ's death), Thursdays (the day he was arrested), Mondays (in memory of the Departed), Saturdays (to honour the Virgin Mary) and, of course, on religious grounds, Sundays. In Lent, Tuesdays and Wednesdays were out too.

Those Spartan attitudes are now out of fashion. Pessaries inserted by women to kill errant sperm were around in Pharaonic times, and condoms of primitive form were used in the Middle Ages. The modern versions are quite dependable, and when used in conjunction with spermicides do the job cheaply and well. Such chemicals dissolve sperm membranes in much the same way as washing-up liquid attacks the grease on plates. Unfortunately they increase the risk of infection by the human immunodeficiency virus. A simple substance, potassium iodide, has now proved effective as a sperm-killer and may replace them.

The discomforts of mechanical contraception (together with its occasional failures) have led to a search for new methods to put a stop to seminal fluid. Popular as the condom remains, there are several new and hopeful ways to interfere with the genital pipework.

A break in the link between testis and outside world is a crude but effective block to the passage of sperm. Vasectomy takes just ten minutes or so, and India's surgical champions can do the operation in forty-five seconds (they gained their expertise at the time of the Emergency of the 1970s, when democracy was suspended and eight million were cut without too much fuss about consent). The Chinese inject superglue into the appropriate spot, which is still quicker, and experiments with microwaves to cook the crucial tube may speed up the process even more. Fifty million men have now been operated upon and in some places the procedure is common. In New Zealand a quarter of all those of reproductive age have experienced it, with the US figure at one in ten.

After the tube has been severed, fifty ejaculations or so are needed before the system is free of generative fluid. Careful months must pass to avoid what surgeons call 'blow-out', bursts through the partly healed wound. The lurking sperm problem, as they call it, is responsible for the few failures of this method of birth control.

Vasectomy can, in principle, be reversed and with almost half of all Western marriages ending in divorce, often followed by remarriage, there is much demand for that option from those whose plans have changed. Success is not guaranteed, even if the removable plugs now used by some clinics make it easier. Some who hope for reversal face a problem: they have made antibodies against their own sperm. The immune system is blind to the presence of such cells, which (concealed within the testis as they are) never normally come into contact with blood. When the two liquids mix, the body identifies the male cells as foreign, and attacks. Even without surgery, thousands are infertile as a result, and their sperm is rejected by their partners on every step of its journey. Attempts to wash off the stain of imperfection always fail. Vasectomy is almost certain to involve a meeting of blood and sperm, and loss of fertility may follow.

Whatever its future in this world of serial monogamy, the operation has an unexpected past, in which – as so often –

the myths of manhood come into contact with science (or pseudoscience). Vasectomy has reinvented itself. For many years, it was used not to diminish sexual function, but in the hope that it would do the opposite. The elderly turned to it when other aids to potency, from wigs onwards, had failed.

W. B. Yeats was among them. In his sixties he wrote with much bitterness about his physical problems and his continued infatuation with young women; of 'Decrepit Age that has been tied to me/As to a dog's tail'. A decade later his song had changed: 'You think it horrible that lust and rage/Should dance attendance upon my old age/They were not such a plague when I was young;/What else have I to spur me into song?'

Yeats owed the lusty and enraged restoration of his final years to hydraulic engineering. The Viennese professor Eugen Steinach had in 1918 come up with the notion that male function (but not, of course, fertility) could be restored with a quick cut through the vas deferens. This, he thought, increased blood flow to the testes and allowed the penis to leap refreshed into action. The Steinach operation ('the general rejuvenating effect has been sufficiently observed both in animals and man to remove any doubt') was soon popular. 'To Steinach' became a verb, and in Vienna itself a hundred university professors submitted themselves to treatment. They included Sigmund Freud, who tried it as a cure for his cancer, with no success.

The fate of a Mr Wilson (who died the day before his planned lecture at the Albert Hall entitled 'How I was Made Twenty Years Younger') rather took the shine off the idea, but Yeats underwent the procedure in 1934, five years before his own demise. The poet was delighted with its results (although he became known to Dubliners as 'the gland old man' as a result). What might he have written with the help of a small blue pill?

CHAPTER 5

MAN MUTILATED

'In the Hereafter, Abraham will sit at the entrance of *Gehinnom* and will not allow the circumcised Israelite to descend into it. As for those who sinned unduly, what does he do to them? He removes the foreskins of children who had died before circumcision, places it upon them and sends them down to *Gehinnom*.' As their sacred writings make clear, to Jews the prepuce is a symbol of rejection. Their identity is not complete until it has been removed.

NORM disagrees, as do BUFF and RECAP. The National Organization of Restoring Men, Brothers United for Future Foreskins, and RECover A Penis all trust in the Tugger – a traction device for prepuces – over the Talmud. To them, circumcision is no more than ritual child abuse. They strive to regain their birthright with weights, tapes and elastic bands which, with patience (it can take five years), restore a semblance of penile normality. Some appliances are more ingenious: as NORM points out, 'The mouthpieces of a number of the larger brass musical instruments are remarkably well suited for modification and may be worn comfortably within the developing foreskin. Tuba, trombone and Sousaphone are among the mouthpieces suggested.' Such ideas are not new, for they descend from the Roman *judeum pondum*, a heavy copper tube placed around the male organ.

Theology hints that Jesus, too, regained his foreskin at the

Ascension. Had he failed to do so, the Saved would themselves have to be operated upon in Paradise so as not to be
more perfect than their Saviour. The prepuce of Christ was
a relic so holy that it was on display in a dozen churches and
was used by St Teresa of Avila as a wedding ring. The last
example did not disappear until 1983.

For Jews the removal of the foreskin is a rite of passage
and a true sign of the covenant and in ancient times even
infants who died at birth were subject to it. Other societies
are adamant that a presence, rather than an absence, is the
true badge of affinity. The Greeks abhorred the whole idea,
because it mutilated the human form. A long prepuce was
much appreciated in their homoerotic society and athletes
were happy to perform almost naked as long as they could
wear a strategically tied string or *kynodesme* (a dog-leash; the
penis was the *kyon*, or dog) to keep the glans hidden. The
Romans were even less keen on the procedure. Mothers who
allowed their children to be operated upon were to be garotted
and hanged on crosses, their dead infants strung about their
necks as a terrible warning.

Saul, as a test of David's suitability as a son-in-law, asked
for the foreskins of a hundred Philistines (and received twice
as many as a statement of just how suitable his daughter's
intended really was). Because the Bible avoids as much as it
can any mention of the male organs themselves, the text is
often taken to refer to a victor's triumphant removal of a
defeated enemy's entire genitalia. A thousand years before Saul,
the Egyptians invaded what is now Libya – and a carved relief
in Thebes shows a pile of phalli displayed before the king.
Such treatment excluded its victims from any hope of salvation: 'He that is wounded in the stones, or hath his privy
member cut off, shall not enter the congregation of the Lord.'

Some psychiatrists see a direct tie between the fate of the
ancient Libyans and that of modern boys. The penis gets five
hundred mentions in the works of Freud and the testicles a

mere dozen, but he was certain that their loss sprang from the same source. In the famous case of Little Hans (which developed the idea of the Oedipus complex and of a child's anxiety about the fate of his genitals) Freud claimed that 'The castration complex is the deepest root of antisemitism as, even in the nursery, the little boy has heard that Jews cut something on the penis – he thinks, a piece of penis – and this gives him the right to think of the Jew with contempt.' Other experts saw circumcision as a useful introduction to pain and a break in the maternal bond, both a great help in the hard adult world. The founder of modern sexology, Alfred Kinsey, cut holes in his foreskin for pleasure and, given his frequent attempts to obtain satisfaction by leaping from a chair with a rope tied around his testicles, risked the loss of other body parts as well.

Prominent as it might be in the works of Freud (a keen Darwinist) the foreskin does not appear in *The Descent of Man*, which is strange, as Darwin – as part of his failed attempt to understand inheritance – was convinced that operations carried out on one generation might manifest their effects in the next. It is an odd and apparently trivial appendage whose symbolism is more discussed than is its biology. Even anatomy texts tend to ignore the structure and may dismiss it (without evidence) as a relict organ, rather like male nipples. As Aristotle noticed long ago, the prepuce resembles the eyelid, a fine membrane well supplied with blood and nerves, firm on the outside and moist on the inner surface. It covers a large part of the flaccid penis and in some men extends beyond the tip. When retracted, the inner surface reveals a ridged band, filled with sensory cells rather like those on the tips of the fingers and on the lips.

Like the hand and the mouth, the prepuce is a sense organ. Its sensitive self is more important to enjoyment than is the glans, which is less responsive to touch than is the sole of the foot. Quite what else it might do is not clear, although its cells pump out lots of prostaglandins, and in rats (if not in men) its secretions are so attractive to females that a male who lacks the structure finds it hard to obtain a mate.

In a newborn child, the foreskin is attached and may be difficult to draw back. Later in life it becomes looser, and in most boys the process is complete by the age of three. At the time of erection the sheath withdraws to provide the extra cover needed for the enlarged organ.

All mammals possess a prepuce, but *Homo sapiens* alone has the urge to destroy it. Nemesis comes in various styles, some more heroic than others. Americans go for a radical 'high and tight' option which removes more than half the skin of the penis, while the original Judaic method (used by much of Islam today, but abandoned at the end of biblical times by Jews because their rabbis were concerned that members of their flock were passing for Gentiles) was far less destructive. After the operation the glans thickens and loses some of its already limited sensitivity.

Maimonides, the twelfth-century philosopher and historian of Judaism, favoured circumcision on those very grounds. In his *Guide for the Perplexed*, he described the operation 'as a means to perfect man's moral shortcomings. The bodily injury caused to the organ is exactly that which is desired; it does not interrupt any vital function, nor does it destroy the power of generation. Circumcision simply counteracts excessive lust.' More important, he thought, a wife with an unaltered lover would find it harder to leave him and to return, as was her duty, to her husband. The evidence is full of bias but gives Maimonides some support. One man improved in adulthood described copulation before and after the event as a film made in colour compared to one shot in black and white. Women (albeit a sample recruited through an anti-circumcision newsletter) also claim to prefer the untouched. The altered, they report, have to work harder and their partners find it more difficult to attain orgasm.

The operation has been justified in many ways. It was once recommended as a specific against masturbation. A President of the Royal College of Surgeons of England in the 1890s was quite forthright: 'Clarence was addicted to the secret vice practised among boys. I performed circumcision. He needed the rightful punishment of cutting pains after his illicit pleasures.' As self-abuse is in fact commoner among those who have been so treated the President was wrong. Kellogg (of cornflake fame, who was much exercised by the pastime and invented his cereal as part of a strict regime to stop it) was more radical. He recommended silver wire: 'The prepuce, or foreskin, is drawn forward over the glans, and the needle to which the wire is attached is passed through from one side to the other. After drawing the wire through, the ends are twisted together, and cut off close. It is now impossible for an erection to occur, and the slight irritation thus produced acts as most powerful means of overcoming the disposition to resort to the practice.' Mr Graham, of Graham Crackers, had more confidence in diet and felt that biscuits alone would remove the temptation.

A fifth of the world's men – seven hundred million altogether – have been circumcised and in the time it takes to read this chapter about a thousand more will be added to the global figure. Although they make up but a small part of the total, the habit is best known among Jews. The ceremony – the *bris* – takes place on the eighth day after birth (which is why nuns wish each other a Happy Circumcision on New Year's Day). In these days of advanced neonatal care parents are allowed to delay the event until a week or more after a premature baby leaves the incubator.

Jewish religious rules are strict (and a close reading hints that even their guests should be circumcised). Previously altered converts to Judaism must themselves suffer a stab to the penis, which upsets the Ethiopian Jews who move to Israel and who see themselves as fully Jewish already. Once, the surgery was done with the sharpened fingernail of an expert, the mohel, but now special instruments such as a heated wire, a Winkelman Clamp or a Plastibell inserted beneath the skin before the knife appears are used. Ritual sucking of blood by the mohel has been abandoned.

The Koran is silent on the practice, but Islam too insists upon it. In the Turkish royal court in Ottoman times, ten thousand boys were cut at once, and in some places group circumcision is still common. The procedure reached its peak in the Yemen, with the *salkh*, in which all the skin was removed from an adolescent's penis, and from his abdomen from umbilicus to scrotum (any flinching put paid to his marriage prospects). In Korea the habit did not begin until independence in 1945. It became common at the time of the Korean War with the influx of Americans, who were the nation's role models in both body and mind. Now nine out of ten boys experience it, often as teenagers.

The procedure is found in many societies who have had no contact with the Middle East or its American diaspora. Australian tribesmen once made cuts not just at the foreskin but across the base of the urethra, which caused urine and semen to emerge some inches before their natural exit. The task involves several operations, and in a few groups opens and splays out the entire tube of the urethra (which psychiatrists ascribe to vagina envy but which may instead be more related to a tribe's desire to identify with a kangaroo, which has a bifid organ).

Whatever excuse is used, the fashion is ancient indeed. An image at Saqqara from 2400 BC shows the temple priests hard at work on a duo of young nobles, with an inscription instructing them not to let their subjects struggle or faint until the job was done. Moses himself was adopted into the royal court and remained unaltered. This led to an awkward moment at the time of the Exodus, soon afterwards: 'Then it happened at a stopping place along the way that Yahweh met

[Moses] and tried to kill him. Then Zipporah [Moses' wife] took a piece of flint and cut off her son's foreskin and touched his feet with it, saying, "You are my blood-bridegroom." So Yahweh let him alone.' This enigmatic text hints at a role of the operation as a fertility rite.

Others imagine an older and a darker history. Marks on the bones of our seven-hundred-thousand-year-old ancestors from the Atapuerca caves of northern Spain suggest that they were, perhaps, the victims of cannibals. In those days, a balanced diet involved the finest meat of all. For some of the sixty cultures known to have indulged in the practice, a fellow man was the second most important source of protein. Circumcision and the mass – wine into blood – may each be relics of a ritual anthropophagy that we prefer to forget.

For ritual, as for diet, fashions can change. The operation is still popular in the United States (President Lincoln, after all, referred to his nation as 'God's almost chosen people') but is in decline. From a 90 per cent rate in the 1960s, about half of American boys are now so treated (more among rich than poor, and among whites than blacks). Even so, two prepuces are lost every minute, and so common is the surgery that a third of those who undergo it do not realise what has happened to them (an even larger proportion of their partners fail to appreciate what their bedmates have been through). The procedure is rarer in Britain, at around one in five of the adult population, with the numbers in rapid decline. Most male members of the Royal Family are among the elect but Diana, Princess of Wales, is said to have abandoned the practice when it came to her own children.

Its predominance in the New World comes from science (or pseudoscience). Some doctors took a Darwinian view: the prepuce was a relict organ, and man, in his superiority over animals, no longer needed it. Lewis Sayre, the founder of the *Journal of the American Medical Association*, was a supporter of the nineteenth-century medical notion known as reflex neurosis, in which diseases were blamed on irritation else-where in the body. He became known as the Columbus of the Prepuce after he discovered that a quick cut with a scalpel restored movement to a paralysed boy. Soon after the publication of his paper 'Spinal Anemia with Partial Paralysis and Want of Co-operation from Irritation of the Genital Organs', his method was shown to cure hernia, constipation and brass poisoning. So successful was the Sayre school that its doctrine became part of normal childbirth.

In spite of such absurd claims, there is some evidence of a small health benefit among a minority of boys. Before birth, the prepuce is sealed onto the glans and, as a result, the fore-skin of most newborn babies is hard to withdraw. Parents may become alarmed as the space beneath fills up into a balloon shape as their baby urinates, but this soon resolves itself. In about one in a hundred the non-retractile state persists to adulthood. It makes personal hygiene more difficult and may lead to infection. Nobody cuts off their ears to avoid the build-up of wax, but to remove the prepuce does solve the penile problem (it can also be cured with a mild steroid cream).

Cancer of the penis affects about one unaltered man in a hundred thousand but is rarer among the circumcised. At least a hundred thousand operations would be needed to prevent a single case (which in any event is almost certain to affect an already aged individual). Each death from penile cancer is matched by two hundred and fifty from cancer of the ovary or breast. Many girl babies at risk of those diseases could be diagnosed at birth with a genetic test and their prospects improved by surgery, but nobody would accept the removal of a child's breasts or ovaries on such grounds. Even women of thirty who have had children are advised about the risks and possible benefits before they decide whether or not to undergo the operation.

Young boys, it seems, do not merit such concern and the surgery is done without much thought for their feelings. The medical decision to do so is made by people who are not doctors, and some physicians feel that the level of pain involved is higher than would be tolerated by any adult. The children do not enjoy their experience, for after the *bris* is over, blood pressure and heart rate stay high and normal patterns of sleep do not return for several days. When it comes to vaccination a few months later, boys who have been operated upon are much less tolerant of the needle than are those for whom the injection is their first experience of the medical world.

Everywhere, surgeons (most of all those who are themselves circumcised) are keener on the practice than are physicians. Fewer than half of British general practitioners would recommend such treatment for a non-retractile foreskin, but nine out of ten surgeons are happy to do so. In the United States the procedure costs three hundred dollars for a few minutes' work and remains popular among those who undertake it. The attachment of their Canadian fellows to such surgery plummeted as soon as the insurance system withdrew payment, and on this side of the Atlantic the National Health Service now only pays in places with many immigrants, where boys treated for religious reasons might otherwise be exposed to amateur operators.

The procedure is not without danger. In the nineteenth century several epidemics of syphilis and tuberculosis were traced to infected mohels as they sucked blood from the wound. Infection is still a problem in the millions of operations carried out under unhygienic conditions in the less developed world. In Turkey, for example (where barbers often do the job), gangrene is sometimes a side-effect. Even after expert treatment, the urethra may narrow because its blood supply is interrupted and the urine emerges in a fine and painful spray. Damage caused by overenthusiastic carving is also a risk. It can lead to an erection curved because of a shortage of skin, or may generate flaps of tissue between the glans and the penile shaft.

As doctors often say, surgery begets surgery; and the adage applies below the waistline as much as elsewhere. In one famous case, identical twin boys were subjected to the treatment. One of them – John – was badly damaged and he was brought up as a girl, Joan. For a time she was happy, but two decades on decided on a second operation, married and adopted children. Large sums have been paid to boys who suffer this fate, and circumcisers are more circumspect than once they were.

Over the years, circumcision has been, like bloodletting, a treatment in search of a disease. The United Nations Convention on the Rights of the Child sets out to put an end to practices harmful to their health. It has been adopted by all nations except the United States and Somalia. The Swedes have passed a law to restrict such assaults on the young. Any medical treatment, they say, calls for informed consent, and if the operation is necessary on social grounds, why not wait until the child is old enough to make up its own mind? Nobody is born Jewish or Islamic; rather they have Jewish or Islamic parents and, at least in some places, are free to decide whether to follow in their parents' beliefs. And what right does one generation have to determine the erogenous habits of the next? Most people in the developed world abhor the mutilation of girls (once justified in the terms still used against boys), but their brothers are attacked almost without question.

The American Academy of Pediatrics, once keen amputators, has weakened its advice; mothers and fathers should, it says, determine what is in the best interests of the child. Today, no national medical organisation in the world is in favour. Only a quarter of unaltered fathers submit their sons to treatment and even those who have themselves undergone the procedure are less keen than once they were. A minor operation is surgery done on somebody else; but a large majority of American women still claim to find a reduced organ preferable on aesthetic grounds and their nation's noble tradition may survive.

Part of North America's obsession with the knife derives from dubious nineteenth-century medicine, but some comes from a study of venereal disease in Canada forty years ago, in which the circumcised were found to be less likely to be infected than those who retain their foreskins. The result was clear enough, but, as in later surveys, the researchers failed to notice that the two groups may also have differed in their sexual habits because Canadian Jews, in general, were rather more faithful than their Christian kin. This confuses the study's interpretation.

AIDS has brought the foreskin back to prominence, and to controversy. In a small study of Kenyan truck-drivers who use prostitutes, the circumcised were eight times less likely to become infected than were men from tribes who keep their prepuces. Childhood circumcision could explain, some say, why infection rates in the Philippines and Bangladesh (where the practice is widespread) are well below those of Thailand and Cambodia, where such surgery is rare.

Persuasive as the figures appear (and they are much publicised by surgeons), several factors confuse the issue. The use of astringent herbs to dry the vagina is common among some of the African peoples who do not circumcise, and this might increase the transmission rate. In addition, the rite is a Muslim habit, and Islamic culture, with its calls for fidelity, may help to slow the passage of the virus.

In some places, indeed, the data point in the opposite direction. Tanzania and Rwanda (where the procedure is common) have a great deal of HIV infection – and the United States, with the highest incidence of the illness in the developed world, has by far the highest rate of such operations. Some surveys even hint that the altered are more, rather than less, susceptible to disorders such as herpes. Most physicians now dismiss the whole idea of a fit between genital surgery and venereal infection.

Whatever the truth, many people believe that such treatment may protect against sexual disease. Private clinics have sprung up in Kenya (where more than half of the unaltered say that they would, given the chance, prefer to lose their foreskins). There and elsewhere, those without a prepuce are seen – in error – as safe from those diseases. Such beliefs increase the rate of spread of the human immunodeficiency virus as the local prostitutes are readier to accept unprotected intercourse with such clients.

Whether to protect against brass poisoning or AIDS, the medical case for removal of the prepuce is unproved at best and those who do it must realise that they do so for reasons symbolic rather than scientific.

Its metaphysical cousin, castration, itself began as a symbol, but went much further. To Aristotle, the testes were no more than weights to keep the seminal passages straight as the essence of manhood flowed in from the blood. A simple experiment proved him wrong. In 1849 the German biologist A. A. Berthold transplanted a pair of testicles into a cockerel whose own had been removed when it was young. Most birds in such a predicament lack the bright feathers and comb present in adult male chickens, but as Berthold reported of his patient, 'So far as voice, sexual urge, belligerence, and growth of comb and wattle are concerned, the bird remained a true cockerel', firm proof that the animal had lacked a signal from the testicles. Orchidectomy, as the operation is sometimes called (orchid bulbs look a bit like testes), is now much used to study the links between the genitals and the rest of the body.

The procedure began long before Berthold. Ancient writers believed, as did Aristotle, that some animals did the job for themselves: 'The beaver, to escape the hunter, bites off his testicles or stones' (sadly for the beaver, the hunters were in fact after the organ for castor oil and not the testes). Men started to emasculate animals around nine thousand years ago, when ploughs first appeared. Bulls will not cooperate to draw such implements unless they have been so treated, and orchidectomy may have begun soon after agriculture itself. It might be much older, for the first Europeans to encounter African Bushmen noted their habit of biting off a son's right testicle in the belief that it made him a better hunter.

The earliest farmers themselves probably suffered it. The first descriptions of human castration insist on the use of a flint knife, a relic of the Stone Age. In ancient times, the operation was brutal indeed. A cord was tied around the testes – and sometimes the penis too – and amputation went ahead. What remained was cauterised with hot irons or tar. The victim was kept thirsty until his wound was in part healed, and then given vast quantities of water until the urge to urinate caused liquid to burst through the scar. Many died, and a castrated slave was several times more expensive than an intact one. The high death rate among such valuable beasts of burden may have led to the replacement of the major mutilation with circumcision. The Old Testament itself is against castration (even for animals) except as a punishment. It associates the practice with the alien lands of Assyria and Persia, the homes of the first farmers.

Scythians (who had – as Hippocrates noted – genital problems of their own) castrated stallions to make them easier to handle and to reduce the numbers of battles for mates. Horses today are treated in the same way. Bulls are cut to make them safer, to increase the rate of growth and to produce more tender meat; pigs to remove boar taint, a stench of pheromone in the carcass (nowadays, taint-free breeds have saved some from their fate). The cannibal Carib Indians of the New World also treated their captives to improve their taste. To remove the testes of household pets makes them less of a nuisance than usual, with dogs no longer anxious to hold a territory and cats less impelled to mark it with a noxious spray. For most farm animals, a tight elastic band – an elastrator – around the scrotum is used, or a burdizzo, a machine to crush the young testicle. For pets, surgery is preferred. A new vaccine able to stop the production of testosterone at source may soon avoid any need for direct attack.

The removal of the testes often acquires a religious tinge, flavoured by a hatred of sex, the work of the devil. Mystics mutilated themselves, and Syrian followers of the Goddess Cybele ran through the streets with their sundered appendages and hurled them through the nearest window. Her consort, Attis, was supposed to have excised his own genitals and, on his feast day, 24 March (my own birthday, as it happens), the ritual went on. The British Museum has a bronze Cybelean castration clamp found in the Thames, and the Roman site of her adherents is buried in the foundations of the Basilica of St Peter's. With the spread of Christianity some of the shrines were dedicated to the Virgin Mary and Christianity itself retained a certain interest in the operation, for Christ himself, in his desexualised state, has some of the attributes of Attis. Hugh, Bishop of Lincoln, who died in AD 1200, was helped by an angel who cut off his manhood to relieve him of impure desires. He followed a noble tradition, for terms related to circumcision and emasculation appear more than a hundred times in the King James Bible.

Castration kept its tie with the Church long after it had died out elsewhere. Emasculated singers first appeared in the sixteenth century when the authorities insisted on the removal of females from choirs, in homage to St Paul's insistence that women should keep silent while in church. In time, they turned against the idea, although castrati were not finally banned until 1902, with a papal decree that 'Those singers who, let us say, are "imperfect" on the physical plane, are totally excluded from the Sistine chapel.' Such songsters continued to appear on stage. In eighteenth-century Italy, several thousand boys a year suffered in the hope of a career. The most famous, Farinelli, was greeted with shouts of 'Hail to the Gelding!' whenever he performed. The last, Alessandro Moreschi, who died in 1922, made the sole recording of an operatic castrato in full song (but alas, somewhat out of tune). Many pieces written for them survive, Mozart's motet *Exsultate, Jubilate* included. Nowadays 'falsetti', who are trained to use their higher registers, take such parts, but a few men with an inborn deficiency of testosterone (and an unbroken voice to match) have made careers as what are, in effect, natural castrati.

To remove the testes is a great restorative, in several ways. In India, eunuchs have stood for election on the platform that their childless state makes them less liable to be corrupt. If the operation is carried out soon enough, it also prevents other masculine misfortunes, from acne to baldness. In the Chinese court, eunuchs were known to be long-lived (the last in the Imperial Court survived to ninety-three) and their Western counterparts do just as well. In the United States in the 1930s, orchidectomy was used, without much thought, as a treatment – or punishment – for masturbation or for minor crime. Forty years later, a follow-up of such unfortunates still resident in mental homes found that, on average, they lived thirteen years longer than their unaltered fellows.

Thirteen years is a long time; after all, to smoke twenty cigarettes a day reduces life expectancy by an average of just five. It hints at the potency of the testicles. The marsupial mouse makes the case for the animal world. In spring, males become territorial and fight for their patch of ground and for a mate. So stressed do they become that all die before their partners give birth. Emasculate them young and they survive for another year or more – which, to a mouse, is an extra lifetime.

Orchidectomy also cheers up the immune system. A rat so treated mounts a more spirited defence against parasites than does his unaltered sib. In the same way, intact reindeer are attacked more by warble flies – nasty creatures whose larvae burrow into the skin – than are those who lack testes. Certain strains of mice carry genes that render them safe from mouse-pox, a lethal illness. Only females and castrated males gain any protection, as the secretions from the testes put a stop to the defensive mechanism. Masculine hormones, it seems, depress the body's police force.

In birds the brain is ruled from below. Remove the testes of a young Japanese quail and its pre-optic area (five times bigger in males than in their partners) fails to develop and his sex life comes to an end. A dose of the essence of testis solves the problem. In winter, some species suffer an effective loss of male function as their testes shut down. Their erotic interests disappear until the next spring.

Many mammals have given up their gonads in the name of science. In most, the prostate gland and seminal vesicles shrink and the structure of the brain changes. Rats so treated increase the synthesis of nitric oxide (the target of Viagra) in the hypothalamus, and behave more like females as a result. Sometimes, such animals lose all interest in mating. Castrated dogs, for example, do not respond to a bitch on heat. Rhesus monkeys retain their desire but become more choosy, with attraction to certain partners but not others, whereas rats are bemused by the whole experience and an altered male continues to mount his mate long after he has abandoned any attempt at insertion.

Their world has some messages for the Bishop of Lincoln and his successors. Hildegard of Bingen (better known for her sacred music) had, like many nuns, a deep interest in erotic matters. She shared a common delusion: that those who lose their testes also lose 'the virile wind that erects the stem to its full strength. The stem cannot become erect to plough woman like soil . . . as a plough is unable to dig up soil if it is without a ploughshare.' In fact the ties between castration and the 'virile wind' are far from simple. Some men manipulated in adulthood do abandon copulation at once, but others take years to do so. Even those who give up sexual intercourse may retain their nocturnal erections, and about a fifth of the emasculated continue with their normal erotic life (without, of course, any chance of children).

However unpredictable its effects on performance, the operation has long been in discreet service to deal with offenders against the codes that control desire. In Germany, its use began with a Nazi statute of 1933, but the Danes had passed a similar law aimed at sex offenders four years earlier. There it continued (with a patient's agreement and in parallel with psychotherapy) until the 1970s. Many of the wrong-doers much improved their behaviour. Such treatment could never become mandatory today as it violates the European Convention on Human Rights on a variety of grounds, from the liberty of the person to the prohibition of torture.

In the late nineteenth century, the President of the American Association for the Study of Feeble-mindedness called for compulsory castration of 'the scum and dregs of mankind . . . degenerates, imbeciles, defective delinquents and epileptics . . . ever with sexual impulses exaggerated'. Another study blamed rape on the 'primitive impulses of the black race' and came up with a progressive remedy, the total abla-tion of the genital organs. A century later, the California Department of Corrections agreed, and in 1996 incorporated chemical castration into its ingenious range of punishments. As Statute AB 3339 says, 'any person guilty of a first convic-tion of specified sex offenses, where the victim is under thir-teen years of age, may be required to receive medroxy progesterone acetate treatment upon parole, and any person convicted of two such offenses must receive the treatment during parole'. A somewhat sinister addendum states that 'If a person voluntarily undergoes a permanent surgical treat-ment . . . he shall not be subject to this section.' The drug is to be continued for an indefinite period and can be forced upon a thirteen-year-old boy who sleeps with a girl of twelve.

The statute followed hysteria about sexual predators: men who pursue the young. Half a million children are molested each year in the United States and, some say, a sixth of Americans suffer such a fate. It costs two billion dollars a year to imprison the culprits, and a variety of penalties are applied against them after release. In Louisiana, they must carry sandwich boards with a list of their crimes, and in Oregon are obliged to place a letter M (for molester) in their windows. Some are forced to carry global positioning system receivers, to track them at all times. A few states keep open registers of offenders and many men have been attacked, or driven from their homes. In Britain a campaign by the gutter press also led to persecution (in part directed at paediatricians, whose vocation sounds a bit like that practised by paedophiles) and, in one case, to murder.

Several states have followed California. The Montana law provides for surgical castration of rapists after a first offence, and Texas asks for the same, albeit with consent, on the grounds that surgery is cheaper than chemical treatment. A prisoner in Georgia has been told that his testes must be removed before he can be released. The sentence is under appeal.

California alone, with a rate of imprisonment per head seven times higher than Britain, releases ten thousand such offenders each year, all of whom might be liable to the drug. Some lawyers are outraged: is this a penalty or a therapy – and, if the latter, why is it compulsory? The American Civil Liberties Union has called the law barbaric. The American Medical Association, too, opposes enforced treatment as a condition of release. Such criminals, the doctors point out, fall into distinct groups. Some deny their acts, some confess but blame pressures such as drugs and some show violent rather than sexual behaviour. A few are paraphiliacs, people aroused by pain and humiliation who direct their attention to children. They are often desperate to escape from their obsession, but are unable to do so. Those rare offenders might be helped by the chemical, but a violent individual is more liable to be driven to terrible rage.

In fact, the drastic penalty set out in the new laws has scarcely as yet been applied (although drug therapy with the agreement of the offender involved is sometimes used). In 1997 a California judge gave a miscreant the choice of five years' suppression of his hormones, or ten years in prison. The defendant chose the former. The sentence (made on the judge's initiative and not on the basis of the 1996 law) was much criticised, and the same lawyer, who had considered it in a second case, instead gave the guilty party twenty-four years inside.

The Golden State's remedy for rapists is better known as depo-provera. It was developed as an injectable contraceptive for females but in the 1980s was discovered to reduce the libido of their partners. It restricts the release of hormones from the pituitary gland and damps down the furnace of desire. Depo-provera does its anti-masculine job less well than does castration, but can reduce the erotic drive and may remove the ability to have an erection. Its side-effects of depression, diabetes and the growth of breasts are judged by the State of California to be unimportant.

Whatever the ethics of forcible medical treatment, it can work. The average American child molester commits more than three hundred offences during his career, and the rate of recidivism is high. More than half of those released from prison return to their fixation. Some psychiatrists claim dramatic success with depo-provera, with a mere one in twenty of those treated still stuck with the habit. The figure is over-optimistic, and the need for an injection a week drives some patients away, but most physicians who prescribe it see a real improvement in their clients' behaviour. However, the approach does not always succeed. A Virginia criminal who had avoided jail by accepting a course of treatment was sentenced to forty years after a later rape of a five-year-old girl.

A new drug might have helped him. Triptorelin is a molecule shaped rather like gonadotropin-releasing hormone, the protein close to the centre of control of the reproductive machinery. It much reduces the level of male hormones and causes a great drop in sex drive. A trial in Israel had dramatic effects. Every patient enrolled in the programme reported the complete departure of his fantasies, from fifty a week to zero. The rate of masturbation fell, as did the exhibitionism, voyeurism and frotteurism which got them into trouble in the first place. The participants were not obliged to continue with the treatment, but almost all did so and remained free of symptoms. Those who did not soon reverted to their previous selves.

Chemical castration, despite strong hints of its use for punishment in the United States, may be a way forward for persistent child abusers. It may – unfortunately – also have a wider role.

Prostate cancer kills almost ten thousand Britons each year and, for Western men, is the second most important form of lethal cancer. Some families are at particular risk, for certain patients carry a damaged gene that predisposes them to the illness. Its normal job is to slow down cell division but the mutated version applies the molecular brakes less firmly and the cells run out of control.

Such malignant acceleration is helped by testosterone. Certain drugs can help, but the most effective way to reduce the effects of the hormone is to cut it off at source. Orchidectomy is often recommended to patients in an advanced stage of the disease. For a sixty-five-year-old in this predicament, the operation adds, on average, seven and a half years of life, compared to just under seven for drug therapy. Some men fear the procedure, but it has fewer side-effects than most of the drugs – which are, after all, in effect a form of chemical castration. In time, the cancer may return in a new form which is impossible to treat.

Orchidectomy might be good for health (or for society) but is unlikely to return to general favour. Western men have become more attached to their genitals in recent years and even circumcision is less fashionable than it was. The Chinese, too, missed their absent members. Sun Yao-tin, the last Imperial eunuch, died in 1996. He kept his testicles in a box to be buried with him and was distraught when the relics were confiscated by the Red Guards, as he feared reincarnation as a dog.

To interfere with such organs diminishes the masculine frame, but parts of the male apparatus have remarkable powers of recovery. The millions of foreskins discarded each year could save myriad lives.

The prepuce has the useful property of almost infinite expansion, once removed. It can as a result be used to repair damage to its owner caused by burns or by inborn deformity. What is more, a baby's foreskin placed in a nutrient solution grows to make a sheet of tissue, which, because it comes from a child whose immune system is not yet mature, is accepted by people in need of a skin transplant.

Two types of cell, one from below the skin and the other on the surface, are used. Each is seeded onto a preparation of cow tendon, or onto a synthetic polymer. Soon they proliferate and, after a couple of weeks, the artificial skin is ready for use. The sheet of material is laid onto the burn. At once it helps to stop the loss of liquid, and in time the patient's own cells move in. In three months, the damage is almost healed. There is almost no limit to how large the expanse can grow and the potential of a single prepuce is measured out in football fields.

That modest structure has been much used to help those with damaged bodies. It assists not only burn victims but diabetics, who may suffer from ulcers which can, if not treated, lead to amputation. Before the new material was invented, a transplant of their own skin (which, for people with severe burns, was not feasible), or of tissue from a corpse, was needed; but now medicine has the potential of an unlimited supply.

Tissue engineering is now big business. It may, some day, replace livers and even brains, but it began with the humble foreskin. The removal of that strip of flesh was once seen by some as an impious attack on the body (in ancient times clipped coins were called 'circumcised' – with a strong hint as to which segment of society was to blame). It may have evolved from an even greater mutilation of the human frame which came in turn from a desire to unman an enemy or from disgust at man's sexual and unspiritual self. Such views are not dead. Florida has an amateur and anonymous gelder who offers castration to those who would, for their own reasons, like to be reduced. He claims to have done the job on scores of volunteers. The gelder charges no fee.

Some still see orchidectomy as a step on the road to salvation and the members of the Heaven's Gate cult who committed mass suicide in 1997 in an attempt to travel to comet Hale-Bopp were found to have undergone the operation. The services of those able to carry it out were once in great demand. The word 'eunuch' appears five hundred times in the works of the early fathers, and today's arguments between the Church's Catholic and Orthodox branches about the need for priests to stay unmarried turn on opposed interpretations of the phrase 'who made themselves eunuchs for the Kingdom of Heaven's sake'. A Russian Orthodox group, the Skoptsy (who did the job with a red-hot knife), attracted a hundred thousand adherents in the nineteenth century. Perhaps they took the Good Book too literally, but clerics of many denominations continued to argue about whether the sacred text meant only what Darwin called 'the senseless practice of celibacy', or more – or (as some hoped) less?

The twelfth-century French philosopher Peter Abelard married, in secret (and after the birth of their child), Héloise, the daughter of a Church grandee. Her outraged father sent a gang of thugs in revenge. As Abelard later wrote, 'They took the cruellest and most shameful revenge . . . they cut off those parts of my body by which I committed what they complained about. They then took to flight, but two who could be caught were deprived of their eyes and genitals.'

The philosopher saw his fate (if not that of his assailants) as a punishment for lust, and became a monk. In his *Dialogue of a Philosopher with a Jew and a Christian* Peter Abelard disputed – as a eunuch – the view that to disfigure the human frame was to condemn it. The body, he argued, means less than the mind within; the castrated, the circumcised and the otherwise damaged are not diminished in the eyes of the Lord. His philosophy was an important step in the slow liberalisation of Christianity which followed.

Peter Abelard tried to escape from the idea of the genitals as a spiritual mutilation. The notion of sex as sin had begun long before, with Gnosticism, a creed born in the Fertile Crescent, the birthplace of agriculture (and of castration), and the site of the Garden of Eden, which to Gnostics was a dream induced in the minds of Adam and Eve by a devil called Ialdoboath. After a period of sublime hermaphroditism, males and females appeared, and God instructed his subjects to realise that the cause of death is love.

The universe, the Gnostics said, was created by demons, and sex was a 'bondage of corruption, a living death, a tomb you carry around with you, a robber who lives in your house who by the things it loves hates you'. Followers of the faith convinced themselves that self-castration was a defence against the infernal fires. Remnants of their belief – older than either Jews or Gnostics – persist in man's eccentric desire to slice off parts of his reproductive apparatus. Today's ability to use the discarded fragments of his organs of lust to save those who face real flames in the world of the flesh retains, as a result, a certain irony.

CHAPTER 6

BOIS-REGARD'S WORMS

Every time a man has sex, he produces enough sperm to fertilise every woman in Europe. He makes two thousand billion of those potent packages in his lifetime, but for the typical Westerner fewer than two succeed. Why are so many called and so few chosen – and how do they do it? Each travels, in terms of its own length, the distance from London to Edinburgh, and each must traverse a series of barriers built by the female before it has any hope of victory. At each attempt only a few hundred reach the egg and just one gets in. What are these remarkable objects, and what is the secret of their success?

The Dutch microscopist Antoni van Leeuwenhoek was, in 1677, the first man to see his contribution to the next generation. He wrote to the President of the Royal Society: 'What I investigate is only what, without sinfully defiling myself, remains as a residue after conjugal coitus. If your Lordship should consider that these observations may disgust or scandalize, regard them as private and publish or destroy them as your Lordship thinks fit.' The aristocrats were broad-minded and allowed his discovery to be made public (although the tale was confused for a time by the question of whether the cells contained small men with moustaches who may or may not have had souls).

Sperm, it seemed, came in two flavours. Twenty-five years

after van Leeuwenhoek, the first book on parasites (*An Account of the Breeding of Worms in Human Bodies* by the French biologist Nicolas Andry de Bois-Regard) pointed out that, as well as the various tapeworms (the 'flesh-eaters') that plague us, 'Man, and all other Animals, comes of a Worm; if the Worm be Male, it produces a Male; and if it be Female, it produces a Female . . . when it is grown to a certain measure, it forces the Membrane of the egg, and then is born.'

Not all such cells are worms. Worms themselves go in for amoeba-like structures, some birds build spirals, insects often make spheres, and crabs prefer a star-like object which explodes onto the egg. They may come in ones or twos or billions, and may travel alone or in gangs in which but a few contain any DNA. Sometimes they work as a team, and in a certain opossum are fused at the head, to swim together until almost at their goal, when one gives up the ghost in favour of its partner. Even among mammals such things vary in length by a dozen times, from the minute sperm of the porcupine to the far longer version of the honey possum. Quite why, nobody knows, as there is no fit between their size and that of the body, although some need to swim less than an inch and others several feet. Even elephants are unexceptional in their ejaculates. Eggs are, in general, eggs; but sperm, like those who make them, are diverse and inventive in their sexual strategies.

Why are they so abundant? Again, the answer is not altogether clear. Sometimes the vast numbers form an army ready to struggle against the donations of an earlier swain. Absence makes the testis fonder, for a husband doubles his production when he meets his wife after a long period away, perhaps as the remnant of a mechanism to flood out the donation of a second male who may have visited in his absence. Self-defence is not the only explanation for masculine excess, as creatures that mate but once also make plenty of sperm. Sometimes a partner gains from her swain's donation and females of certain fruit flies who receive many ejaculates lay more eggs than do those who must be content with a few. A preference for a munificent mate could then force her suitors to make more and more of the generative fluid if they wish to succeed.

Genetics also plays a part in sperm's profusion, for many are deficient in some way. Each represents a new mixture of the DNA of its maker and some draw bad hands in life's poker game. A tenth of man's germ cells bear a chromosome error of some kind, and almost all carry new mutations, some of which are harmful.

Most of the travellers on the rocky road to the next generation abandon hope on the first leg of the journey. Great numbers of our own sperm fail to mature, and about a third are abnormal (as a result, a man makes the same number of normal sperm as does a hamster). Some have a swollen head, or a tail unable to lash in the correct way. In chimpanzees, in contrast, almost all male cells have at least the appearance of good health. Why men are so feeble, nobody knows. If so many of their cells are unable to complete their journey, van Leeuwenhoek's horde may be less potent than it seems.

Compared to eggs, a male's tiny cells seem cheap and are scattered with what looks like careless abandon, but they are less of a bargain than they appear. In India, loss of semen is said to lead to dry skin, anxiety, painful joints, palpitations, headaches, swollen gums and bad breath. A tablespoon of the magic substance demands the nutriment contained in a hundred pounds of food. The ancient Greeks felt the same: they referred to it as *stagon encephalou* – a drop of the brain – and blamed the blindness of the Sybarites on their abundant ejaculates.

Such claims are inflated, but are not altogether wrong. A small worm much used by geneticists avoids the sex problem for much of the time as it lives as a hermaphrodite and needs to make just a couple of hundred sperm in its lifetime. Now and again, a male is born – and such individuals produce thousands of such cells. The extra burden shortens their lives by a quarter. Other creatures also find the job expensive. Adders emerge from hibernation in the spring and the males bask for weeks in the sunshine without eating. First the testes enlarge as sperm is made; then the snakes find a partner, fight off the opposition, court and copulate. Each task uses up reserves and each involves a substantial loss of weight. To make sperm costs as much in terms of solid flesh as do the later and more visible events. The emperor penguin, indeed, saves its ejaculate up for an annual emission.

Some mammals too have spermless seasons and the walrus bothers for a mere three months a year (for men, the seventh-century Chinese document *The Principles of Nurturing Life* recommended a twice-weekly springtime ejaculation, but complete winter abstinence). Even so, mammals manage on a sexual shoestring in comparison to snakes, for less than a hundredth of their energies are devoted to the job, compared to a fifth directed to muscles or the brain.

When desire triumphs, performance may suffer. As Galen may have said, '*triste est omne animale post coitum, praeter mulierem gallumque*': every animal is sad after coitus, except the girl and the rooster. Information on women and poultry is patchy at best, but for males he was right. A domestic sheep forced to ejaculate eight times a day with an electrical probe inserted into its anus suffers a drastic drop in sperm numbers. Its wild cousins on the Hebridean island of St Kilda fight with much bitterness over mates. Many are killed as they butt each other, but the winners copulate a dozen times each day. A dominant animal's sperm count drops as the season wears on and, although he mates with the same eagerness as before, he is firing blanks. Humans reach that sad state after a mere half-dozen ejaculations in twenty-four hours.

Male cells demand investment because they are hard to make. They are built to travel great distances, and their DNA goes through some complex and expensive processing before it is packed on board. The sperm's cargo of genes is not a simple duplicate of those of its maker. Almost every cell in the body has two copies of each chromosome and each gene, one from each parent. As a result, the DNA in each new sperm (and egg) must be reduced by half. In addition, the contents of such cells are reordered by recombination. All this involves a special form of division known as meiosis, a term derived from the literary device of using understatement to make a point ('I will not be sorry when I have finished this book').

Meiosis, the formative moment in every man's history, involves a doubling of his DNA followed by two cell divisions. In the first, the genes are reshuffled into new mixtures. After that solitary sexual event, the second division produces structures ready to mature into sperm.

Such cells are the last members of a long and privileged lineage. In the first weeks of embryonic life, a group of special cells (the germ line, as it is called) is set aside and kept insulated from contact with the rest of the body. At puberty, those bearers of the genetical holy grail (and they are rare indeed, with the fate of the next generation entrusted to a few tens of thousands of cells, far fewer than in a drop of blood) begin to divide. Until the last moment they copy themselves in the ordinary way, but are allowed a special dispensation. In most cells, the tips of chromosomes fray each time they divide, with the loss of a small sector of DNA. In the germ line, an enzyme repairs the damage and rejuvenates the family line.

Then meiosis begins. It gives rise to round cells that, in spite of the revolution they have experienced, look much like their forerunners. In time, these biological aristocrats mature into sperm.

All the cells of an early embryo have the potential to develop into a variety of tissues: muscle, blood, nerve and so on. Such stem cells, as they are called, lose their adaptability as they settle into their adult role. The germ line is different:

the precursors of sperm stay forever young (which, given the new interest in the use of stem cells to grow tissues for medical use, may make testicles rather helpful to their owners even when old age has put paid to their primary function).

A male cell goes through much modification before it reaches adulthood. Maturity takes around two months and happens in waves along each testis tube. First, it becomes longer and develops a head, a solid structure ready to burrow into its target. Soon a tail and a midpiece appear. The midpiece is packed with mitochondria – molecular batteries for the genetic vehicle – while the tail contains special proteins that ratchet over each other to make it lash back and forth, together with a set of molecular springs to help in the energetic task.

Each human sperm must travel more than a foot – hundreds of thousands of times its own length – through each partner's tract before it reaches its destination. On the way it is bathed in a variety of fluids able to help or hinder its passage.

Just a small part of the ejaculate comes from the testes. The rest is added in the treatment factories that feed into the genital tract. In the epidydimis, a storage vessel just beyond the testis itself, sugars (a useful fuel) together with a variety of proteins are added and the solution is concentrated by a hundred times. Here sperm learn to swim. After ten days or so they are pumped into the vas deferens – the tube between testis and outside world – and gain a protective dose of vitamin C. The prostate also plays a part as it adds prostaglandins and other substances. These cause smooth muscle to contract and may assist the cells on their way once inside the other party as they persuade her to squeeze them towards their target.

After certain negotiations between the partners, semen appears on the scene. The first spurt is filled with sperm, while the second is richer in sugars. Within a few minutes the liquid coagulates, with the help of proteins from the seminal vesicle. In some mammals the congealed material forms a tough plug that denies entry to others. An hour later the jelly turns back to fluid. Much of what passes from one participant to the other leaks out, but millions of cells fight their way onwards.

In the Third Reich, brothels collected semen from unwitting customers with a plan to employ it as a substitute for serum, to be used in transfusions. The notion is quite mistaken, because sperm carry a series of potent antigens recognised as alien by those who receive it. A female responds to the sudden arrival of a foreign tissue in her genital tract in much the same way as she would to a kidney transplant. Her immune system is summoned up, and white cells of the blood turn upon the invaders. They may, perhaps, remove sperm unable to meet some criterion of her own. As a result, very few make it as far as the egg – just one in a million for men (and pigs, who ejaculate two thousand times more than we do, are even less successful). Of those few hundred weary voyagers, a solitary cell reaches its destination.

The road to the egg's DNA narrows in several places and in some creatures is decorated with blind turns, each a trap for any traveller unwary enough to be diverted. Most of the cells are held immobile on the walls of the recipient's tract and go no further. A few are altered by secretions which strip off their protective coat, and swim on. Each step of the way, the female keeps a careful eye on who is allowed to proceed. A mere one in ten is allowed to take the sacrament and to lose its coat. For two hours or so after its purifying experience a sperm can respond to a message sent from the liquid around the egg and is able to move towards it. The myriad that do not receive the molecular seal of approval are doomed.

Most of man's most precious cells live for just a day or so inside his partner (a few survive for up to a week, which is useful in rape cases). Other creatures are more persistent. Bats copulate in the autumn and hold sperm over the winter, to let it free as spring returns. If a fresh swain appears in time, fine. If not, she has a reservoir for emergency use. Insects are even more thrifty, and the queens of certain ants can store such cells for several years.

In the end, and for however long delayed, all sperm must face the final test. A simple experiment shows how stringent it can be. A white male rabbit was mated with a partner of the same colour. After a few hours, the surviving sperm were flushed from his mate, just before they reached the egg. The few dozen rescued cells were then inserted into a second female who had already been mated with a brown animal. Ten million sperm were introduced by the second, brown, individual, and a tiny fraction of that number by the scientists – but some of the young were white. Somehow, the trial by ordeal had sorted out the white rabbit's finest cells and allowed them to prevail over the as yet untested selection inserted by the second mate.

Insects also make choices based on genes. Females are anxious to avoid sex with a close relative and have a security system to keep incestuous sperm at bay. Crickets inseminated with a mixture of sperm from their brothers and from an unrelated male use only that of the latter, as further evidence of the power of the internal filter.

The few cells allowed to approach their goal face the most rigorous examination of all. The test set at fertilisation takes less than two hours to complete, but has several papers – and but a single candidate can pass. First the egg has a series of vestments which must be unwrapped before entry is allowed. The outermost is crude but effective, as it excludes cells from the wrong species. When the jelly is stripped off in the laboratory, human sperm are able even to penetrate hamster eggs (but not to fertilise them).

Only part of the egg is permeable to sperm, and those who make their attack in the wrong place are doomed. The cells lucky enough to land in the target area then face a set of molecules on the egg surface to which they must fuse. Some are members of the ADAM family of proteins, much involved in interactions among cells elsewhere in the body, and, oddly enough, also found in snake venom. Each sperm carries several other types of attachment molecule which zigzag back and forth across the cell membrane. The egg has a matching set, to which the victor affixes itself.

ADAM has other unexpected relatives. The most complex family of human genes, with more than a thousand members, expresses itself in the nose (which is why we are able to sniff out fine wines, sewage and the like). It counts ADAM among its members, which makes sense, for plenty of creatures – dogs included – are most interested in the scent of potential mates. A few of the genes involved in personal odour are at work as identity molecules in sperm and egg, and a certain sniffing out of the best may take place in a more intimate fashion.

Males must keep up to date in the sexual race if they are to be accepted. Many sea creatures release sperm in billions into the sea; in Samoa, indeed, the locals scoop up and eat the seething mass of sex cells from palolo worms, who emit them just once a year. In marine snails such as abalones, which also emit such things in huge numbers, the recognition molecules evolve with remarkable speed. For abalones and all other creatures, the egg has interests of its own, because if two sperm get in, it dies at once. Given the numbers available, it has plenty of choice about who is allowed to enter. The egg changes its locks and the sperm is forced to follow. Abalone eggs have a coat of protein chain-mail, which unravels when the victorious sperm arrives, under the influence of a male protein called lysin. The lysin gene is among the speediest to evolve of all, with a rate of change among relatives fifty times greater than that of any gene in mammals. Mammalian sperm have a rather similar mechanism, and the protein involved also shifts at a great rate. All this hints at the ferocity of the race for acceptance.

At last the victor's head fuses to the egg surface and, with the help of a special enzyme and its beating tail, it fights its way through and its contents are pumped inside. They include DNA, of course, but the double helix is not alone. It is accompanied by a special messenger that persuades the egg to complete its own maturation, and an enzyme able to make the universal signal, nitric oxide. At once the gas causes a wave of calcium ions to pass across the egg surface. This renders it impassable to all other entrants and the solitary and triumphant genetic message is safe from dilution.

Several other tasks remain before the marriage can be consummated. Women carry a lot of baggage, which has interests of its own. Eggs transmit not just their own DNA but that of mitochondria. In ancient times, those structures were bacteria and a few of their ancestral genes remain. Like their relatives in the great world outside, they struggle to make a livelihood. They do not like outsiders – foreign mitochondria included – who might chip away at their assets. The cell's own genes are also suspicious of their domesticated helpers who might, given a chance, revert to their aggressive habits. To avoid a battle between maternal and paternal types, and to prevent sex between different lineages that could generate new and hostile forms, evolution has come up with a compromise. Mitochondria and their genes stay as monopolies, segregated into females, who alone pass them on.

In human sperm the structures are marked with a molecular brand and are at once destroyed. A certain fruit fly makes an enormous sperm. Its mitochondria survive within the egg, but the larva builds its guts around them. In Freudian style it gets rid of its father's contribution in its first defecation.

If a conflict between the two sets of genes does break out it can stop fertilisation dead. The ingenious owner of an American fertility clinic solved his patients' problems with an injection of egg material from a third party into the expectant cell. His sperm was welcomed, but the procedure led to the birth of babies with – in effect – three parents: mother, father and the donor of the extra mitochondria. What this might do to the child, time will tell.

The egg does not complete its own meiosis until well after a sperm has entered and the cell has begun to divide. The union of each parent's nuclear DNA takes even longer.

Sperm influence the egg's prospects in other ways. Fathers put molecular marks on their genes. Without them the embryo cannot develop properly (which is why cloning is so hard to do). Their imprint is wiped clean in the next generation. The male also contributes structures that help the fertilised cell to divide, together with a cellular scaffold which sets the embryo on the path to adulthood. The point of entry determines where the first cell division takes place. Descendants of cells near the site of arrival split earlier than others and develop into the embryo itself, while cells born further from the impact site make the secondary membranes and blood vessels around the fetus which help to nourish it.

Male cells have another crucial role. They are the source of most of the genetical damage faced by each generation – and, as a result, of the raw material of evolution. Sperm have a harder time than most cells. As they swim through the oxygen-free environment of each partner's genital tract, they make toxic byproducts and injure their own DNA. Male cells are so fragile that the children of heavy smokers are at a higher risk of cancer, perhaps because spermatic genes are harmed by the poisons inhaled.

Whatever insults it may suffer from the external world, a sperm's main cause of damage comes from within and is, alas, impossible to avoid. Men make their reproductive cells all the time, from puberty to death. A woman, in contrast, has her one true sexual experience – meiosis – before she is born. Her germ cells enter the process while she is still a fetus and then take a long sabbatical, which does not end until the egg is released. As a result, every egg goes through the same number of cell divisions while within its owner, however old she might be. In contrast, men never rest, with a fresh division in the germ line every two weeks or so. The DNA in each sperm goes through round after round of multiplication – which leads to a far higher incidence of error among fathers than mothers.

Age plays a crucial part. The sperm made by a twenty-year-old is separated by about four hundred cell divisions from the sperm that made him, while those of a man of fifty have gone through more than a thousand extra rounds of replication. Because of the chances of a mistake each time the DNA is copied, old roués have far more mutations than do young bucks. For an elderly gentleman who hopes for parenthood, the news is bad. Fifty-year-olds emit twice as many sperm with extra Y chromosomes than do men of half their age and, for certain genes, the figures are even worse. In some conditions (such as the common form of dwarfism) the children of a father of advanced age are at twenty times higher risk than are those of a younger male.

The noble families of Europe speak of the power of the years. An aristocratic daughter gets one of her two X chromosomes from her father, while a son obtains his sole copy from his mother. Among seven hundred such families, daughters with fathers of fifty or more died on average at seventy-four, while those with fathers of less than thirty lived for an additional three years. The age of the father had no effect on the life expectancy of sons (who inherit from him the Y and not the X). The feebleness of aged men has led the American Fertility Society to accept only sperm donors of under fifty, with the British even more cautious, with a limit of thirty-five.

Mutation does at least mean that men do something useful, as they are the source of the raw material of genetic change. Some of the errors are harmful, but others do good and are soon picked up by natural selection. As a result, a lot of evolution takes place in the male line. The effect is best seen on the X chromosome. Because females have two Xs and their partners just one, the structure spends, over the generations, half the time in males compared to the other party. As proof of the masculine role, the build-up of mutations is slower on that chromosome than elsewhere.

Some mutations kill off their carriers, but others damage the reproductive cells. A sperm contains more than a thousand different proteins, and is bathed in a fluid filled with many more. In fruit flies, a tenth of the vast numbers of mutations known, whose main effects extend from changes in eye colour to shifts in wing shape or behaviour, damage reproduction. A variety of inborn human illnesses, with symptoms as distinct as deafness and diabetes, do the same. If the fly figures apply to ourselves, mistakes in five thousand genes might harm a man's machinery.

Those faced with such a quandary often turn to medicine for help. Infertility clinics put cells through their paces with computer-aided analysis of movement, with a barrier of cow mucus as a trial of their ability to make the journey, or with the denuded eggs of hamsters in a test of their power to penetrate the target. Sometimes a husband's sperm can, after a self-induced ejaculation or minor surgery, be used to fertilise his wife. If even that fails, there are always plenty of other possibilities. In one clinic, a tenth of the women who came for help became pregnant even as their partners stayed sterile. They had taken their destiny into their own hands.

More often, a less informal arrangement is called for. Sperm donation has been around since the Scottish anatomist John Hunter impregnated the wife of an impotent draper with her husband's own cells two centuries ago, but did not come to general attention until 1909. In that year, the American physician Addison Davis Hard claimed to have carried out the first donor insemination twenty-five years earlier, while he was a student at Jefferson Medical College in Pennsylvania. His patient was a Philadelphia merchant, and the semen came from 'the best-looking member of his class' (who, most of the audience assumed, was Dr Hard himself). The lady involved was unaware of what had happened as she was under anaesthetic at the time. After Hard's lecture, she was at once identified, and there was an uproar.

Thirty-five years later, a British doctor admitted that she too was involved in the practice. At once the establishment, in the form of a commission chaired by the Archbishop of Canterbury, was called in. Sperm donation, it declared, was a sin equivalent to adultery and should be a criminal offence (the Pope agreed and called for all the parties involved to be imprisoned). A 1960 committee was more sympathetic, but was concerned that 'sperm donation is an activity which might be expected to attract more than the usual proportion of psychopaths' and which should be discouraged. Not until 1982 did the law accept that the habit was widespread and should be legalised. Now tens of thousands of babies are born into the world each year in this way.

In France, only men who already have children are allowed to donate, but Britain prefers medical students. It has a panel of around three thousand regular contributors. They are not easy to find. Some potential donors give up when they realise what is expected – fertility tests, a genetic screen, then half a dozen visits before the first sample and twenty more to produce the maximum allowed (to prevent cheating, they are not allowed to work at home).

In the early days, some British donors had two hundred progeny, but now each is allowed to father a maximum of ten. In Denmark (with a population a tenth the size of that of the British Isles and with an equivalent increase in the chance that children of the same father might meet and marry) he can have more than twice as many. The Cryos Company in Aarhus ('We keep the stork busy!') is the world's largest bank, with hundreds of suppliers. It often exports to

Scotland, where the material is in particularly short supply.

The United States allows couples in search of sperm to shop around. Most centres screen their contributors for a dozen or more genetic diseases, and for all the common venereal infections. Then comes a search through the panel for a match based on ethnic background, eye colour and the like. What was the donor's most memorable childhood experience? ('I was actually thrown into the water at age five. It was literally sink or swim. To my surprise and the wonderment of others involved, I could actually swim.') What was the funniest thing ever to happen to you? ('I was involved in a swimming race. Around 800 yards, my swimming trunks fell off. I finished the race to a standing ovation. It was the largest crowd to watch what is otherwise a boring, tedious event.') The aquatic individual on offer is six feet tall with a Ph.D., and admires honesty, humour, compassion and objectivity – and, to clinch matters, his grandparents died at 81, 85, 93 and 94. His sperm sells at three hundred dollars a shot.

Some donors have a more tormented history. In 1980 a sample of sperm was taken from a young American braindead after a car crash in the hope of using the cells to fertilise his wife's eggs. Another young man bequeathed his frozen sperm to his girlfriend and then committed suicide. The British are stricter about the rules. A wife whose husband was dying of meningitis asked for semen to be removed while he was in his final coma. It was, but as written consent had not been given the authorities refused to allow it to be used. The wife went to Belgium where she gained a court's agreement. Her son was born in 1998 and a second child four years later.

On the other side of the Atlantic, too, the law has not come to terms with the problem. An attempt to donate sperm to their girlfriends by two Virginia prisoners under sentence of death was described by the Governor as 'brazen' and by a court as 'frivolous'. The application was refused and they were killed, but in 2001 an equivalent request in California was granted (the dissenting judge described the claim as 'ill-conceived'). The latest scandal involves Mafia mobsters who have tried to smuggle their sperm from prison. The police have seized some specimens as fruits of a crime and have applied to the courts to destroy them as contraband. The situation remains fluid.

Some soldiers now freeze samples of their sperm before going into battle. Radiation and chemotherapy kill germ cells rather than their owners, and cancer patients, too, may wish to store their own sperm. Several have done so (even if in practice rather few use the material). Tissue from the testis can be frozen and replaced after the treatment, when it goes back to work. A few unfortunate children also develop the disease. Half are cured (which means that about twenty thousand British adults have survived such an illness) but most of them are sterile. As children, needless to say, they cannot make sperm, and some doctors now freeze their immature cells in the hope that some day they can be persuaded to develop further. Others hope to graft a boy's testis into his father's and to return it after treatment is over.

Wherever the cells come from, artificial insemination is common, and succeeds for about three-quarters of the couples who try it. The process is simple enough: sperm enters a partner's tract more or less as normal, but without benefit of penis. It usually makes the first step of the journey from the donor's organ in frozen form. The Italian biologist Lazaro Spallanzani found in 1776 that sperm packed in snow could swim again after half an hour of suspended animation (nowadays dry ice or liquid nitrogen is used). As ice crystals kill the cells, an antifreeze is added. Bull sperm can be kept alive for forty years and might be stored almost indefinitely. As mutations might build up, ten years is seen as the safe limit for ourselves.

Sometimes the approach fails and sperm is obliged to meet egg in a test tube. *In vitro* fertilisation has moved on since the birth of Louise Brown in 1978, with a quarter of a million annual attempts in Europe. The Danes are enthusiasts, with one baby in twenty conceived by some form of assisted reproduction.

For those with severe damage to the sperm, even *in vitro* fertilisation is not much help as their cells cannot penetrate the egg. Intra-cytoplasmic sperm injection – ICSI, for short – inserts the masculine contribution (a mature but incapable sperm or its immature predecessor) into an egg with a needle a tenth the size of a hair. After the lucky DNA has been forced through the female's defences, fertilisation proceeds as normal.

More than a hundred thousand children have been born through ICSI, with a hundred or so centres in the United Kingdom alone. The procedure is not without risk. A husband sterile because of damage to his Y chromosome may pass the problem to his sons. When faced with this dilemma, some couples ask only for girls. Worse, the process might itself harm the embryo's prospects. Some husbands fail to pass on the Y chromosome at all – which causes severe damage to any egg that receives such a sperm – and others are infertile because of genetic errors coded for on other chromosomes, which may themselves be transmitted. As the point of entry in a natural fertilisation lays out the body plan, to puncture a random hole could be a mistake. A few of the babies born in this way have suffered from errors in chromosome number or from other genetic diseases, perhaps because unqualified sperm have been forced through the fertilisation exam. Even so, most of the children appear to be quite healthy.

When even such heroic treatment fails, a donor may seem the only option. Some husbands are unhappy with the whole idea. How is it possible to bond with a child born of somebody else's genes? Science offers them hope. Mouse testis cells can be moved from an infertile strain into another able to reproduce, or into a rat. The cells develop into sperm, and the sterile mice then pass on their heritage via a proxy. Mice have not yet been persuaded to nurture human sperm, but perhaps cells from an infertile man could be transferred into a laboratory culture from a testis no longer needed by its owner, to grow to maturity. A healthy – and generous – individual might even be able to process a friend's faulty sperm in his own testis and to deliver it when called for, so that the barren husband becomes the true parent. It may some day become possible to take body cells, halve the number of chromosomes and use them to fertilise an egg. This would cut out the testicles (and perhaps their owners) altogether.

As is the case for the anti-impotence industry, the vast sums spent by men anxious to deliver sperm – any sperm – to their partners hides a larger truth: that for most of the time males are keen to avoid the penalty levied by nature as a tax on pleasure. One couple in three depends on a male-based method of contraception, from condoms and premature with-drawal to a judicious cut through the tubes. Simple and effective as such methods are, some people dislike such crude technology. The new biology gives them hope.

Van Leeuwenhoek, the first to see his magical cells, wondered whether he should reveal his discovery as 'the world, which is coarse and vicious enough, might use the knowledge of nature for its own ruin and increasingly debauch itself in depravity'. He meant, of course, birth control. Science has allowed depravity to reach new heights.

In the 1960s the female contraceptive pill transformed men's lives. In these more egalitarian days, the search is on for a male equivalent. Eight out of ten British and American men would, they say, be happy to use it and, perhaps more remarkably, an even larger proportion of their partners would trust them to do so.

The task will not be easy, for man's insistent cells – unlike those of his opposite number – are copied all the time and delivered in enormous quantities. Sperm are resolute little objects, and none can be let through as even a count reduced to a hundredth of normal gives a pregnancy rate too high to be acceptable.

They can be suppressed in several ways. High levels of testosterone itself can close down the hormone factory and the hormone has been tested as a contraceptive, but its side-effects of acne, high blood pressure and rage make it unsuitable for general use. A modified form developed by the World Health Organization is as good at the job as is the Pill, and a single injection can stop sperm production for a year (for reasons unknown it sterilises Asians better than Europeans). As steroids may damage the liver and might have other long-term effects, the prospects of testosterone-based birth control are uncertain. Trials have also begun of synthetic progesterone capsules implanted into young men, which block sperm production.

It might be safer to interfere further up the line of command, in the brain. The hormones of the female cycle are much involved in male function, and a pill based on them might put a stop to sperm as it does to eggs (even if the men involved must also take masculine substances to avoid the growth of breasts). Other drugs more associated with wives than husbands may also help. The 'morning-after pill' sometimes turned to in an emergency stops sperm dead, but its effects are too unpleasant to allow regular use.

Various other chemicals have been tested. The Chinese worked for years on gossypol, an extract of cotton seeds, which puts a stop to the crucial cells. Unfortunately its effects are sometimes permanent. They have turned their attention to a wild plant, the *lei gong teng*, or thunder god vine, used against arthritis in traditional medicine.

Contraceptive design has moved beyond the God of Thunder. A drug once used to treat high blood pressure blocks channels in the cell membrane through which calcium ions pass. An unwelcome side-effect is sterility. The company involved kept its potential quiet because of a fear of lawsuits by those who wanted to control blood pressure rather than family size, but now the special channel that pumps ions across the sperm membrane has been found. If it is obstructed, the sperm cannot swim. The protein responsible could become a target for a male birth-control pill – and the drug might also be taken by women, to halt any errant cells within their own bodies, with the first unisex contraceptive.

The new genetics offers other ways to stop unwanted sperm. Infertility can be caused by antibodies and it may be possible to inject a man, or his partner, with his own germ cells (or with the chemical cues on their surface). Mice engineered to lack a certain receptor within the cell make no sperm at all and a drug aimed at the molecule might work well. And why not target the machinery of meiosis, unique as the process is to germ-cell formation?

Such notions are still in the future, but other forms of spermatic technology are already well advanced. Such cells are, of their nature, no more than delivery packages for genes. Biochemists have spent years in search of transfer systems of their own, based on viruses into which alien genes are spliced and which carry their cargo to the target, or on tiny golden bullets coated with DNA. They have given sheep and cows human genes that allow them to make proteins such as growth hormone in their milk.

Now industry has become interested in nature's emissary, sperm itself. A protein from jellyfish that shines green when exposed to ultraviolet light was the first to make the trip. Its gene was inserted into a virus and the mixture added to mouse sperm. These were injected into eggs – and several of the young glowed in the dark. Once adult, they passed the gene onto their own progeny, conceived in the normal way. Rhesus monkeys are less cooperative (the protein is made in the embryo but not in the adult), but spermatic engineering may soon become common. The latest twist puts genes into sperm stem cells, which may open the gates to the next generation in a way natural enough to persuade it to accept alien genes more readily than now.

Genetic engineers are also attracted by the ejaculate. Milk is not an ideal vehicle in which to make foreign proteins because animals make it only from time to time and because it has many complicated ingredients of its own which must be cleaned out before the product can be used. Semen is a better candidate. Boars make a pint a shot, and do so on demand. Already, mice with an added gene for human growth hormone plus an on–off switch from a testis protein have been persuaded to manufacture the material in their semen. If pigs could do the same, they might, with some help from their keepers, produce large amounts of such substances. Filter off the sperm and a pure protein remains. Some pig farms already have two thousand boars busy at sperm donation. A single such enterprise could make enough growth hormone to satisfy the world.

To some, all this suggests that science has gone mad; but, for sperm, simplicity still has a place. Indian biologists have found that a three-week course of one-hour soaks in a hot bath causes six months of sterility, and have improved on this with a cup of water heated by microwave into which the relevant parts are suspended until all sperm have been stopped. Elegant as such techniques might be, van Leeuwenhoek would scarcely have approved.

CHAPTER 7

BEND SINISTER

Dial, if you are in the United States, 1-888-RU-MY-KID (or if that makes your intentions too obvious, try a competitor, from 1-888-HOME-DNA to 1-800-DNA-TYPE). Britain is more discreet with its telephone numbers, but a dozen or so services offer paternity testing. A quarter of a million Americans a year turn to such companies and an undisclosed number try the transatlantic equivalents. Some are men, desperate for a ruse to escape the cost of child support, but most are women concerned to prove that a male who denies paternity is in fact the father. The businesses involved tend to emphasise peace of mind over paternity suits, but in about three-quarters of their cases the alleged father is, deny it as he might, in fact the biological parent.

According to Alexandre Dumas, 'The chains of matrimony are heavy, and it takes two to bear them, sometimes three.' Bastardy has, as a result, been around as long as marriage itself. To Freud, a belief in fatherhood marked the emergence of abstract thought and the origin of society. Paternity, after all, is based on deduction while motherhood makes itself obvious at once. If a husband's faith in his child's parentage is misplaced then the social order itself is at risk. The penalties for illegitimacy – to women, to children and sometimes even to men – are as a result severe. The accused tend to deny all knowledge of their acts and once they could get away with it; but now science proves how often males tell falsehoods to females – and, too often, to themselves.

The ancient name for the penis, the 'yard', was an inflated statement of its length, and that for the testes – the 'stones' – came from hyperbole about their weight (and led to leaden medieval jokes about men made 'lighter by two stone'). Bearers of the *SRY* gene tend to boast about sex as much as they lie about fatherhood, which makes the truth about paternity hard to establish. Most people believe, without much thought, that bastardy arises because men are of their nature less faithful than women. The claim cannot be true, for reasons of simple arithmetic. It takes two to tango, and the average number of encounters must be the same for each sex.

Even in the face of such stern statistical reality, people may differ in achievement one from the other. If in a village of ten men and ten women, every inhabitant stays in a loyal marriage, then the mean number of partners for each person is (needless to say) one and there is no chance of illegitimate children. If, on the other hand, a single libidinous individual has an affair with all the females, with his rivals left out in the celibate cold, the average number of copulations for either sex remains at one, but the variation in success is far greater for males than for their opposite numbers and most children will be born to a temporary union. For those who believe in biology as destiny, such behaviour is no more than human nature, as the carriers of Y chromosomes are programmed to be promiscuous and to invest as little as possible in the results of their excesses.

The Descent of Man makes great play with the idea that such differences in sexual success led to the evolution of beauty and even to racial divergence: 'unconscious selection would come into play through the more powerful and leading men preferring certain women to others' (Darwin does admit that 'very ugly, although rich men, have been known to fail in getting wives'). The claim that the wealthy have more mistresses is ancient indeed. The *jus primae noctis* – the right of a master to all young girls – goes back to the Epic of Gilgamesh, but the idea fizzled out into the symbolic privilege to rest a naked buttock on the bridal bed. Royalty has always done well, and George IV was described as 'rather too fond of woman and wine' at seventeen. He persuaded each of his lovers to donate a lock of hair, and by his death had enough to stuff a modest sofa. Don Juan, with a thousand and three conquests in Spain alone, was based on a certain Don Miguel de Mañara, a libertine who, after a vision of his own funeral, became Brother Superior of the Spanish Brothers of Charity. With typical masculine hyperbole, his grave is marked: 'here lie the bones of the worst man the world has ever seen'.

Others see the violation of the marriage bed as an ethical, rather than an evolutionary, issue and point at the poor, rather than the rich, as the main culprits. The Scots blamed the 'low moral tone' of farm servants and the English the 'healthy animalism of the country people, and the weakness of safeguards, principle, self-respect and public opinion'. Strict penalties were needed to stop it.

Don Miguel's exploits have faded into legend. They symbolise every man's supposed dream of success: to be affluent, powerful and free to spread his DNA in all directions. His claims, and those of other sexual athletes, rest mainly on anecdote (although the noble Don did have a factotum who kept lists). They were once impossible to test. A mere lack of information has not stopped endless and futile speculation about the evolution of promiscuity, but today the truth can be revealed.

DNA has brought many a blush to male cheeks. In heraldry a bend sinister – a bar from bottom left to top right across a coat of arms – shows illegitimate descent from some noble ancestor. The Dukes of St Albans, of Richmond, of Buccleuch and of Grafton, each a scion of Charles II and one of his several mistresses, have never made a secret of their badge of unofficial parenthood. Now the double helix stands ready to unveil the behaviour of any man, noble or otherwise, and with or without a family tree. It has put a leftward bar across countless escutcheons.

Such ornaments can be expensive. In ancient Rome, an illegitimate child could inherit nothing as it had, in legal terms, no father. Even children of the same mother but different fathers were not defined as kin as they were thought to share no blood. English law had a similarly jaundiced view of parental rights; as the eighteenth-century jurist William Blackstone put it, 'The interests of husband and wife are one – and that one is the husband.'

Friedrich Engels in *The Origin of the Family, Private Property and the State* noticed the tie between the inheritance of genes and of goods. The notion of legitimacy, he wrote, did not appear until the emergence of capital, the medium which allows rich fathers to pass their wealth to the next generation. They want to restrict the gift to those who carry their own biological heritage and invented marriage to do the job. The destitute bastard who began with capitalism would, thought Engels, vanish with the abolition of private property.

Governments are still much involved in the theory of reproduction, but their interests are now centred on cash rather than politics. American states rake in five billion dollars towards the costs of child support each year from errant fathers, and in the United Kingdom the once much-criticised Child Support Agency also acts, under the guise of social service, in part as an agent of the Treasury.

Some states of the Union are punitive, and a mother who wants help must name the supposed father before she can qualify. To test her allegation is simple: compare the genes of the person nominated with those of the child. If the infant has a heritage that could not have come from him, he cannot be to blame. If the child does share variants with the accused, the case remains open as, however powerful the test, it is never possible to be absolutely certain that a specific man is responsible. Genetics can do no more than include a defendant in the group who might be at fault. Even so, as more and more genes enter the equation, the assembly of the blameworthy narrows to such an extent that it becomes unreasonable for the accused to deny his guilt.

'Unreasonable' is a term much debated by the law, but the chances of statistical error are now astronomically small. Astronomy itself accepts at least the possibility of life on Mars, and some innocents will no doubt be found guilty because of a chance match. In addition, biology cannot guard against technical failures (which happen more often than the laboratories like to admit), falsified evidence or eccentric events such as the sudden appearance of a long-lost twin. All this is often pointed out from the dock, but rarely with success.

How useful any variant might be in the hunt for the errant male depends on how frequent it is in the population from which he comes. A label shared by many is not as effective a badge of guilt as is one common to only a few, which means that a supposed parent with a rare inherited illness is less able to deny fatherhood of a child with the same condition than is – say – a blue-eyed man faced with a blue-eyed child. Geography comes in too and a blue-eyed Greek in such a predicament is in a weaker position than a blue-eyed Swede.

But which clues should be used? Some couples quarrel because their son or daughter does not look much like the husband. The explanation is often innocent. Likeness is in the eye of the beholder and is not a reliable guide to kinship. When asked to match photographs of children with those of parents most people get it right just half the time, which would not be much help in court. For those who do not understand the rules, even genetics can be deceptive. Simple Mendelism allows two dark-eyed people to have blue-eyed offspring, as long as each carries a hidden copy of a gene for blue eyes. The intricacies of eye colour, with at least two genes involved, with changes in hue with age, and with doubts about the exact definition of the terms 'blue', 'green' or 'brown' also mean that – contrary to popular belief – two blue-eyed parents may have a child with dark eyes.

The ABO blood groups were used in a paternity case within months of their discovery. They gave the first hint of the power of biology in the courtroom, but they, too, are less clear-cut than they seem. The system has three variants. The *O* gene in effect removes a certain cue of identity on blood cells, while the other two alter its structure. An individual of blood group O bears two copies of the *O* gene. Somebody of blood group A or blood group B may either have a double dose of the *A* or *B* gene, or a single copy of either, matched with an O. People with *A* matched with *B* have the AB blood group. A marriage of two O people can produce only O children, but other crosses are more complicated. Any marriage between, say, an A woman who has an *A* gene matched with *O* and a B individual with his *B* gene matched with an *O* can produce A, B, AB or O offspring.

Once, the law could not cope with such complexity. In 1944 Charlie Chaplin was pursued by the US Federal Bureau of Investigation – whose director, J. Edgar Hoover, saw him as a communist – for white slavery, for transporting a teenager across state lines for immoral purposes. Chaplin had without doubt had an affair with the young actress involved, a certain Joan Barry, but managed to escape from the clutches of the FBI. However, he was later pursued for child support by Joan Barry herself. The paternity test was unambiguous: Chaplin was blood group O, she was A and the baby was B. He could not have been the father, but the jury found him guilty because they decided that the child looked like him. Chaplin was forced to pay seventy-five dollars a week until the boy grew up.

The verdict shows the limits of the law as it tries to cope with science, but odd things happen with ABO that might allow a clever advocate to make a case for Chaplin's guilt. Two distinct genes, coded for in separate places in the DNA, are involved. The substance made by the O variant is a precursor of those coded for by A and B. A certain enzyme, coded for elsewhere in the genome, converts the O material into its A or B products. A very few people lack the crucial catalyst and cannot, as a result, make substances A and B, even if they carry the actual A or B genes. They are, as a result, always group O. If a BB person in this situation marries somebody of group O (who is almost certain to have a working copy of the enzyme), then the child inherits a B gene from one parent and a working copy of the enzyme from the other – and is blood group B. Two O people can as a result have a blood group B child.

That fact has been used by lawyers to bamboozle juries (and once made an appearance in the soap opera *General Hospital*), but the damaged version of the protein is frequent among white-skinned people only in a village on the island of Réunion in the Indian Ocean (which cuts down its usefulness as an alibi). The variant is rather more abundant among the native population of India.

Jesuitical arguments about blood groups have been superseded by technology. The first hint of its power came from the discovery of a series of short repeated sequences of DNA bases that vary in number from person to person. Molecular scissors are used to cut the double helix in specific places, and differences in the number of repeats then reveal themselves in the lengths and numbers of the pieces as they travel through a sieve when exposed to an electric field. Each of us has a unique genetic signature of this kind – a DNA fingerprint, as it is known. The method was first used in criminal cases, but is now much applied to test paternity.

DNA fingerprints are not foolproof, as some of the repeated sections have a mutation rate of as much as one in a hundred. A guilty person might as a result get away with his true paternity because a genetic error which happened while his sperm was made has confused the facts about fatherhood. Nowadays at least nine different repeats are used to minimise the chance of false innocence.

The completion of the map of our gene sequence means that the fingerprint and its relatives will soon be out of date. Individuality has become, in effect, unlimited. A search for single-letter variants in the DNA of a mere two dozen people from across the world found a million and a half variable sites – and that excludes the shift in the short multiple groups of DNA letters. Such single nucleotide polymorphisms, as they are called, mean that the DNA of two individuals differs in millions of ways. As the variants are reshuffled whenever sperm or eggs are made, the number of permutations is far, far more than the number of people alive today, those who have lived or those who ever will live.

Genes make mere hints about false paternity obsolete. Most of the companies who offer a test now claim a chance of less than one in a hundred thousand of a match arising by chance (and their figures are conservative) and few of the accused now try to argue their innocence on the basis of statistics.

Even the dead are not safe from the testimony of the DNA. Larry Hillblom, joint founder of the world's largest shipping company, retired to a tropical island where he pursued his interests in fishing, flying and copulating with adolescent girls. In 1995 his plane crashed and his body was never found. Soon a variety of young mothers from all over the Far East appeared with babes in arms to claim him as the father and, as an incidental, their right to a share of his six-hundred-million-dollar estate. The absence of a corpse made a genetic test difficult (and clothes and brushes that might have carried his genetic fingerprint were discovered mysteriously laundered and buried in his garden), but a check of eight of the many children who claimed descent found that four of them shared a male parent – and their mothers, otherwise strangers, had independent evidence of a link with Hillblom. The children received ninety million dollars each as eloquent proof of the power of the double helix.

The test that made them rich depended on variants scattered across all chromosomes. Because such genes are mixed each generation their history becomes blurred with time. Just one clue allows infidelity to echo undisguised. The Y chromosome is a lasting mark of fatherhood which has put an indelible bend sinister across many a family crest. Its magic can chase adulterers deep into the past. Although it is rather less variable than are other parts of the genome (a search through twenty people uncovered just four hundred differences among them, a fifth the diversity of a typical chromosome), it provides more than enough evidence to track down an errant father, even if he died long ago.

Families are defined not just by shared genes but by a second, non-biological, label. The surname, like the Y, is a statement of affinity to a masculine line (even if few names are four hundred letters long). If males are faithful, names and genes should go together. Any mismatch hints at past adultery (although the story can be confused when, as sometimes happens, a family with no sons asks a husband to take up their surname to preserve their lineage, or when the same name has two distinct origins).

In the North of England, sykes are ditches and the word is often used as a place name (as in the village that has become a suburb of Rochdale in Lancashire). Sykes is in addition a surname first recorded in the thirteenth century in the country around Huddersfield. It began, no doubt, when it was picked up as a badge of identity by some dweller on the banks of such a drain. By chance, one of the ten thousand Britons who has inherited the name is a geneticist. He wrote to a sample of his fellows and obtained DNA from fifty of them.

Half the respondents bore the same Y chromosome while a few others differed in a minor variant which arose by mutation over the past few hundred years. The remainder had a great mix of types, rather than just one (which might have marked an independent origin of the Sykes name around some second northern ditch). Their diversity hints at an ancient influx by genetical black sheep who bestowed DNA upon the family without benefit of a surname. The incidence may seem high, but the name is at least twenty generations old and a simple sum shows that the actual rate of false paternity was little more than one in a hundred per generation.

A fit between names and Y chromosomes is of interest to the police, who may soon be able to track down a criminal who leaves DNA at the scene with a simple glance at the telephone book. Bill Sikes, the brutal thief in *Oliver Twist*, bore a minor mutation of the famous surname and would have problems in a world of DNA technology. He met his own end in an accidental hanging, without the help of forensic science.

As Dickens was well aware, men can be as dishonest in their marital as in their fiscal affairs. However well behaved they are at home, their habits can change when they get the chance (as the waves of rape at the time of the Balkan wars show). The genes of Black Americans are eloquent testimony to the long history of male betrayal.

Most African Americans (the celebrated gospel singer Jubilant Sykes among them) possess not an African but a European surname. Slaves were once seen not to need family names at all, but in time it became customary for them to take up the tag of their master. In revulsion against this false heritage some have replaced a European label with a title of African or Islamic origin (which explains the rebirth of Cassius Clay as Muhammad Ali). Malcolm X, that celebrated figure, discarded his own birth name – Malcolm Little – as a badge of shame, but preferred to use a solitary letter rather than to imagine an African ancestor. Malcolm X's Y would almost certainly have told him a less welcome tale about the past.

Thomas Jefferson was the author of the Declaration of

Independence. Two years after the death of his wife in 1782 he became the United States Ambassador to France. On the trip to Paris, his daughter was accompanied by a slave girl, the fourteen-year-old Sally Hemings. The young Hemings soon became pregnant and had a son, the first of her three boys. An embittered journalist spread the tale that the child was in fact that of Jefferson (who had by then become the third President). He had, the scribbler claimed, 'kept, as his concubine, one of his own slaves'. Jefferson denied the charge, but the story stuck. The poet Thomas Moore much disliked the United States on his 1804 visit. He criticised the behaviour of its head of state: 'When he fled/From the halls of council to his negro shed/Where blest he woos some black Aspasia's grace/And dreams of freedom in his Slave's embrace.'

Two centuries later, the Y chromosomes of several Jeffersonian descendants (in fact those of his father's brother, as Jefferson himself had no direct male link with the present day) and of the progeny of Sally Hemings were compared. To general astonishment, those of the scions of Hemings' youngest boy, Eston (born when Jefferson was sixty-five), were identical to the Y chromosomes of the family of the third President who had, it seemed, been less than faithful to his office. The story is muddied because Thomas Jefferson's brother Randolph (who of course shared his Y – as did two dozen other Virginians of the period) spent much time at Monticello, the President's estate. Randolph often mixed with the slaves – as one recalled, 'he used to come out among the black people, play the fiddle and dance half the night' – and may himself have been the culprit. On the other hand, the records show that Thomas himself, rather than his brother, was always around exactly nine months before the birth of Sally Hemings' children.

Whatever the details (and the story, some say, was spread by supporters of William Jefferson Clinton to put his own private life into context), history and technology combine to

identify an ancient case of uncertain fatherhood. The association of Jeffersonian descendants has still not decided who to
allow into their group. Some insist that many of the claimants
are not related even to Sally Hemings, which may lead to
legal action against false claims of false paternity. Even so,
plenty of hopefuls are still in the queue. Like the descendants
of Charles II, the glamour of a noble ancestor removes the
blemish caused by his ignoble behaviour.

Brazil, when it was discovered by the Portuguese in 1500,
had two million Amerindian inhabitants. Over the years half
a million colonists – most of them male – arrived, together
with large numbers of African slaves. Nowadays the nation
presents itself as a mixed and reasonably open society. In some
ways the genes agree, for they show that most citizens who
identify themselves as white have in fact a large proportion of
Amerindian or African ancestry. Many poor and dark-skinned
Brazilians who are in their own eyes European would not be
recognised as such in Europe itself as, on the average, fewer
than half the genes of such people come from the Old World.

The paternal lineages of those ambiguous Aryans tell a
different tale. Almost every Y chromosome borne by a self-
styled Brazilian white, however dark his skin, did indeed orig-
inate in Europe. The genetic mix was one way: white men
mated with their native or slave paramours, with almost no
traffic in the opposite direction. Today's social tolerance, so
far as it has been achieved, emerges from a history of sexual
inequality. In North America the story is much the same, and
Eston, the supposed son of President Jefferson, himself passed
as white when he grew up. His own children were accepted
into white society, with a European Y chromosome to accom-
pany their African genes.

The reality of racism in the West disguises the depth of
the problem elsewhere. The Indian caste system – which
touches the lives of a billion people – began three thousand
years ago. Castes are classified by origin and by occupation,

into Brahmin (priests), Kshatriya (warriors, the group to which the Buddha belonged), Vysya (traders) and Sudra (who serve the others and are looked down upon as the 'once-born'). A more recent group, the Panchama or untouchables, are at the bottom of the heap. Indians can identify status by dress and speech but also believe that the upper and lower ends of the social scale look different; that rank lies in the genes.

At least where men are concerned, they are right. For the female line (as manifest in mitochondrial DNA) Indians as a whole resemble other Asians, and the castes are not distinct. For the opposite sex the story is quite different. All castes have some Y chromosomes whose ancestors came from Western sources such as Iran, but the upper castes have far more than do those from lower down. The warriors, the Kshatriya, are closest of all to Western males. The class system was brought by invaders from what is now Iran and the lands further towards Europe. They formed an aristocracy which, as in Brazil, impressed its genetic stamp upon the local women. Since then, brides have been allowed to move between strata, but grooms cannot, and the disdain by each caste for those below is such that there is still not much overlap in their paternal lineages.

Ideas of race or class in Brazil, India or anywhere else are, needless to say, not based altogether on biology but turn on the beliefs of the society involved. The notion of bastardy, too, depends on the human construct known as marriage. For legitimacy, the boundaries move fast, and the law may struggle to keep up.

What, for example, should be done if for medical reasons only one parent has a biological tie to the child? When a wife – with the full knowledge of her husband – is impregnated by donor sperm, what is the status of the offspring? The legitimacy of the marriage is unchanged – so why doubt the status of what it produces? The boundary between legitimate illegitimacy of this kind and that frowned on by the state can be hard to define.

In the first days of donor insemination, parents often hid the truth and put the name of the husband on the birth certificate. Anybody foolish enough to follow the letter of the law was obliged to leave the paternity box blank. The parents could confer legitimacy on the child only if they adopted it. If they did not, the marital father had no liability for the infant and could abandon his family with no penalty. In 1982 the Warnock Committee called for a change in British law to allow the husband to appear on the birth document – but with the crucial caveat 'by donation' after his name. Parliament saw this as too painful for the children involved and allowed the marital father to appear unadorned.

Since 1991, when the figures were first collected, over twenty thousand British babies have been born by sperm donation. Attitudes to legitimacy may change, but biology still insists on a say. What if an unknown benefactor should turn out to have an inherited disease, or an insurance company were to ask the youngster, when adult, about illness in the family? Perhaps someone could, without realising, plan to marry a half-sister or half-brother, the issue of the same philanthropist. To guard against such problems, the government keeps, when it can, a secret record of genetic fathers. When the children involved grow up, they have the right to ask whether they were born through donor insemination and whether they are indeed related to somebody they hope to marry. The law allows no more – but by so doing it accepts some genetic basis to identity.

Attitudes vary from place to place. In the United Kingdom, the anonymity of donors is absolute (in spite of recent attempts by some children to use the Human Rights Act to find their fathers). Donors themselves are never informed of the birth of their child and three-quarters of the children born in this way are not let into the secret. Most who are told cope well with the information, but others react with horror and search for an identity distinct from that on their birth certificate.

In some places they can use the law to do so. The United Nations Convention on the Rights of the Child gives everybody the right to know their parents. Society has become more open in most matters, so why not for paternity? Sweden, Austria and a few other countries already allow adolescents born with the help of a third party to learn the truth (although the change in the Swedish regulations led to a drastic drop in the number of donors). American adoption records were until not long ago sealed for ninety-nine years, and most states followed the same rule for sperm donors. Some children resent this. The Bastard Nation group pushed an initiative onto Oregon ballot papers to allow them to discover their birth parents and won the right to do so in 1998. Most agencies remain opposed to the idea, and attempts to spread the movement have stalled.

When it comes to paternity, California is more subtle than most. Both those who give and those who receive sperm can do so with a pledge of anonymity or, if they prefer, with an 'identity release' which allows the child to learn, should he or she wish, who their father was. The policy began in 1983, and in 2002 the first teenager to take advantage of the law met her biological father. Several more such reunions are in the pipeline. Britain remains undecided, but the government is now testing public opinion about whether – for future cases – the rules might be relaxed.

Among the most advanced societies, the whole notion of illegitimacy may be abandoned. The French Revolution decried the idea: 'As if there were some people more "legitimate" than others!' Those born out of wedlock were rather a 'precious human resource of soldiers and mothers', to be known as *enfants de la patrie*. The populace agreed and the proportion of such births rose from one in fifty before the political upheaval of 1789 to five times as many after it.

In Russia some revolutionaries were even more radical. In 1917 the town of Vladimir turned to Engels and *The Origin of the Family*. Its commissars issued a decree: 'From the age of eighteen, every young girl is declared state property. She is obliged, under pain of prosecution and severe punishment, to be registered at a bureau of free love. Men likewise have the right to choose a young girl who has reached the age of eighteen if they are in possession of a certificate confirming that they belong to the proletariat . . . In the interests of the state men have the right to choose women registered at the bureau even without the assent of the latter. The children who are the fruit of this type of cohabitation become the property of the Revolution.'

The idea was popular (at least among men), but did not catch on. The central question, as ever, came from biology: who will cover the costs of childcare? If society will not (and the Revolution showed no interest in the task), who will? And if males no longer need to pay for their erotic diversions, what will limit their excesses? What might the man in the street get up to, given the chance?

His genital appetites are notorious, but exaggerated. Free access to women, revolutionary or not, is an attractive notion, but the evidence that it has ever happened is ambiguous at best. Most cases of false parenthood come not from some universal orgy, but from occasional individual lapses. For the science of sex, like all others, anecdote is not enough. Kinsey had an interviewee who, he claimed, had copulated twice a day, with a variety of partners, for thirty years; but Kinsey took his subject's word for it. Moulay Ismail the Cruel, Sultan of Morocco from 1672 to 1727, is the *Guinness Book of Records*' coital champion. He had, he asserted, eight hundred and eighty-eight children by hundreds of different wives (his female equivalent in the *Guinness* pages had a mere sixty-nine offspring).

Moulay must have been busy (and the lady champion was without doubt making up her figures). Each of his wives was fertile for just a few days in each cycle, each copulation had but a small chance of success and, in his day, half of all children died at or before birth. To justify his claims, Moulay Ismail must have mated several times a day for the whole of his reign (and he still found time to kill thirty thousand Christians with his own hand). Many of his supposed progeny probably came from the efforts of others or made up their allegation. The same, no doubt, was true of Vladimir's radicals: plenty of desire, but rather less performance.

When men – Alawite sultan, Marxist zealot or hungry academic – get together to compare copulatory notes, sex and lies tend to share a bed, which is unfortunate for students of human nature (illegitimacy included), but disastrous for the scientists interested in the spread and control of venereal disease.

The truth matters when it comes to the clap, as a single active individual can do tremendous harm. In the eighteenth century, cures for such diseases took up more newspaper advertising space than did any other product. For the past century or so, the problem has dwindled but the spread of AIDS, syphilis and the like has renewed concerns about masculine behaviour. A new survey of previously undiagnosed cases of gonorrhea found a twentieth of young American men to be infected – which could mean serious illness for their partners.

At the height of the AIDS scare, the British government planned a national survey of coital habits – which much excited those whose attempts to understand the evolution of reproductive behaviour had been slightly marred by a complete lack of data. The idea was quashed on moral grounds by the then Prime Minister, Margaret Thatcher. A medical charity, the Wellcome Foundation, did the job instead with its National Survey of Sexual Attitudes and Lifestyles. At the turn of the millennium, it updated the figures.

The new probe hints that a third of men have had more than ten partners by their forties, and about one in a hundred manages in excess of a hundred conquests. Most, though, were happy with half a dozen or fewer.

Strange things happen in any society, sexual or otherwise, in which most players stay within a small group, but occasional high-flyers mix with new people. It takes just a tiny number of movers and shakers to link the world together and to ensure that individuals who never meet face to face (or at points lower in the anatomy) are closer than they think. Lotharios, rare as they are and inaccurate as their claims might be, much boost the numbers who turn up at the Special Clinic (where business is booming, with a 20 per cent increase in visits by male clients in the past decade). The wanton are a real problem for those who try to model the spread of epidemics, and the urgent need to measure just how promiscuous they might be has given a new insight into the secret world of manhood.

The Scottish Football League is a convenient metaphor for the universe of sex. Almost nobody who plays for that somewhat unsuccessful team Queen of the South, in the Scottish Second Division, will ever receive a ball directly from the feet of a player for the great Turkish club Galatasaray. There has never been a direct transfer of a player from Scotland's Not Quite Finest to Turkey. Even so, a shift of a single team member to, say, the English League and another player's transfer from a different English team to the other side of the Bosphorus puts, in two or three passes, the heroes of Dumfries in honorary contact with the feet (and, through a different route, with the genitals) of the greatest players in the world; Scotland, to England, to Turkey, to the globe. In the same way, a few promiscuous men soon link more-or-less faithful marriages in distant places.

In such networks, most people belong to societies that feel as isolated as Scottish football, but, unknown to their members, all have an indirect but intimate union through a small number of shared links. The pattern allows a small number of orgiasts to start an epidemic (which is why AIDS spread so fast among homosexuals, most of whom were fairly faithful but who were linked by a minority who were anything but).

Embroidery of the truth by aristocrats is to be expected, but a close look at the surveys hints that less illustrious people are also economical with the *actualité*. The problem came from an internal check on the figures. In any accurate census, males and females must on the average (given the two-to-tango rule) report the same number of partners. They do not: men always claim a higher score than does the opposite sex. In the United Kingdom the Wellcome averages were, in the 1980s, in turn 9.9 and 3.4 and in France 11 and 3.3. The new survey narrowed the gap, with just twice as many partners for males, but both sets of data show the same lack of consistency.

Either men exaggerate or the distaff side is too modest, but if the basic facts about the numbers of mates cannot be trusted, what is the point of gathering the data? How are we to count the infectious Don Juans, or track down the unsupportive fathers? And what about the endless claims by those keen on the biological roots of society that every man's behaviour is based on his struggle to join their ranks?

The United States hints at an unexpected truth. The European health detectives had overlooked a small but crucial group. In their vast promiscuity they give a new insight into the rules of fatherhood.

In Lincoln's upright nation, men report less than twice as many encounters as do their partners; the All-American Male is perhaps more honest than his British counterpart (or his third President) but still, it seems, prone to hyperbole. The key lies in the details, in a small and shady group of enthusiasts who are easy to miss in the broad sweep of a general survey. The health detectives concentrated on prostitutes and their clients. In Colorado, a not untypical state, each prostitute had an average of six hundred sexual contacts a year and a few had more than ten thousand. In the huge national surveys no women reported figures even close to that. The sex industry had been missed, in part because it employs so few people (about one woman in five thousand in the United States) and in part because its employees live uncertain lives, hidden from curious eyes. Quiz those few males who claim more than a hundred mates in a lifetime and most admit that their success turns on commerce rather than charm.

A correction for cash-based copulation shows that most people tell the truth – more or less – about their private lives, with almost no disparity in the numbers of partners reported by each sex. The huge number of contacts by prostitutes also reduces the apparent difference in promiscuity between men and women. Illegitimacy, the new figures show, is, most of the time, in the family. It involves an occasional fling by a restless wife or errant husband and is not a microcosm of Moulay Ismail's times, when a few men impregnated vast numbers while their peers stayed glumly celibate.

Some Y chromosomes do, nevertheless, find their way to where they should not go. Attitudes to fatherhood change with class and with time. Legitimacy turns on circumstances, and on how it is defined. Today's Britons may seem less reliable than their stout northern ancestors but just how many children fall outside the category nowadays is hard to pin down. The figure of around one in ten once much quoted by geneticists seems to be no more than an urban myth. Paternity-testing companies and the Child Support Agency come up with even higher proportions, but their clients are a selected group whose marriages are already in trouble. A DNA survey of the general population in Switzerland gave a number ten times smaller. That must – presumably – be multiplied by some factor to reveal the truth for the British.

The new negligence turns on social change (and from a mistaken belief in the power of contraception). It also emerges from women's improved status and their ability to raise children alone. The Pill put them in control of fertility, and more and more choose to have babies without masculine support. As the importance of social position fades, people are more prepared to accept children for what they, rather than their parents, might be.

The state still finds it hard to deal with the nature of fatherhood. Until 1957, the application form for a US visa asked: 'Do you practise, or advocate the practise of, polygamy?' Utah's constitution was firm in its defence of the social norms: plural marriage was a habit known only to savage peoples (which put its own founders outside the civilised pale). Polygamy, a relic of Mormon times, was abandoned when the state wished to join the Union, and the Church President had a vision that told his fellows it was time to return to the federal rules.

The Old Testament itself is full of multiple marriage (even if the 'Adulterer's Bible' of 1805, which printed the Seventh Commandment as 'Thou shalt commit adultery', earned its printer a fine) and the idea of legitimacy is itself rather new. The first demographers scarcely bothered with the concept, but its importance grew with the Industrial Revolution. Thomas Malthus had a robust Anglo-Saxon view of children born on the wrong side of the sheets and unsupported by their father: 'If the parents desert their child, they ought to be made answerable for the crime. The infant is, comparatively speaking, of no value to the society as others will immediately supply its place.' He would be baffled by today's figures.

In the United Kingdom one birth in each dozen or so is registered by the mother alone – a figure twice that of the shameless 1960s – and half of all conceptions take place out of wedlock. The facts of false fatherhood (and of remarriage) mean that hundreds of thousands of children are brought up by men who they assume, in error, to be their biological father. For most of the couples involved, only the ceremonial has been abandoned as the parents – often of bourgeois stock and each keen to help raise the child – choose to live in a union founded on their own desires rather than on a contract approved by the state. In the great national survey of erotic habits carried out in 2000, a sixth of adults under forty were in cohabitation with a long-term partner – about half as many as those who had entered formal wedlock. J. B. Priestley described the marriages of his day as 'a long dull meal, with dessert as the first course'. In these new and epicurean times, serial monogamy – wedlock, divorce and remarriage – has become common, and the obsession with birthright is on the way out.

Even with such changed views, the problem of the absent father remains. Britain has (after Belgium) the second highest divorce rate in Europe. As a result, a fifth of all fathers do not live with their child. Unmarried fathers now have a right of access after a split, but just a few thousand requests are made each year and fewer than half of those who leave keep any contact. The Child Support Agency in the mid-1990s estimated that the same proportion paid nothing at all to support their progeny.

Such behaviour can cause real difficulty for the youngster involved. Oedipus is well known for his relationship with his mother, but he grew up in a fatherless home. Laius conceived him when drunk, and ordered his wife to destroy her infant. She did not and, years later, he took unwitting revenge when he killed his sire in a quarrel.

Many children have suffered the Oedipal problem. In the old days, premature death, not a desire for fresh sexual pastures, put paid to large numbers of male parents. Their absence posed a real as well as a metaphorical threat. The rate of infant mortality doubles in such families (and in Sheffield a century ago, more than half of all fatherless children died in infancy). His loss also increases the likelihood of failure at school, drug abuse, delinquency and suicide.

Now that DNA has made the truth about paternity so easy to discover, the nature of legitimacy – and of fatherhood – needs to change. Some native peoples of South America are in advance of the developed world. Among the Bari of Venezuela, a woman may copulate with several partners, each of whom sees himself as in part the father – and such people live up to their obligations, as a child with several 'parents' has twice the chance of survival than does an infant with just one. In the cities of Brazil, too, a man often provides powdered milk ('father's milk', as the locals call it) which confers legitimacy on a child born to his partner by another.

Remnants of a similar idea survive in the developed world. An ancient tenet of English law holds that a baby born in a marriage is seen as the issue of the husband. The assumption may not always be true, but at least it ensures that the new arrival is looked after. Whatever the faults of its parents, the infant is, in the end, not to blame. To use science to exclude the contribution of an adult to its welfare chips away at children's rights.

The persistence of that rule has led to acrimony in cases in which divorced husbands are forced to support an infant proved not to be their own. DNA evidence is used in the criminal courts to acquit the innocent – why not, they ask, in family law as well? Even so, few American states exempt a man who has helped to bring up a child from penalty if he is shown not to be the biological father. Some legislatures allow the result to be used until the infant is three, after which a husband will be held responsible for his wife's offspring, whatever the DNA might say. The courts almost always hold to the well-tested notion that the welfare of a child has priority over the facts as revealed by science.

All this accepts that to raise a child involves obligations that spread far wider than its parents. However precise the results offered by paternity testers, the truth was recognised by societies that flourished long before they appeared: that fatherhood means more than genes alone.

CHAPTER 8

JAMES JAMES'S SKULL

In 1927 my elderly relative James James sold his head. He got seventy pounds for it, a substantial sum for a skull in those days; particularly as the terms were cash in hand, cranium when available (and delivery was delayed for a decade until his death). The relic can still be seen in the National Museum of Wales in Cardiff, although it has few visitors nowadays.

James James lived on the slopes of Plymlymon, in central Wales (where my own grandfather, James's cousin, was born). There, many believed, survived the aboriginals of these islands, pushed to the edge by an influx of aliens. As the anthropologist H. J. Fleure (who paid for the head) wrote, 'The Celtic fringe is in a sense the ultimate refuge in the far west, wherein persist old thoughts and visions that else had been lost in the world . . . They were a strongly-built people with dark colouring of hair and eyes and perhaps still a dark tinge in the skin, with long, high-ridged heads, big eyebrows and deep-set eyes . . . such people survive on the Plymlymon moorlands . . . The Little Dark People are found nowadays in the rural population as a rather acquiescent element; this type is often a large majority in a religious or bardic gathering.' His notion of the Welsh as an entity was an ancient one. They were the remnants of the Celtic race, once widespread, but driven to the margins by more aggressive nations from the mainland of Europe.

The Welsh themselves disagreed. They were not mere relics, but fearless colonists in their own right. Madog, a prince of Gwynedd and one of seventeen brothers, was born at Dolwyddelan in Snowdonia. To avoid struggles about land, he set off with a band of companions on the good ship *Gwennan Gorn*, into the unknown. They landed in Alabama, where a plaque now records their exploits: 'In memory of Prince Madog, a Welsh explorer, who landed on the shores of Mobile Bay in 1170 and left behind, with the Indians, the Welsh language.'

Robert Southey, no less, wrote an epic poem on Madog's adventures, which included the conversion of the Aztecs to Christianity (it was epic enough to lead Byron to write: 'Oh Southey! Southey! Cease thy varied song!/A bard may chant too often and too long!'). The Welsh, the poet sang, were explorers, equals in spirit to the English or anybody else. As further proof of their intrepid ways, the New World itself was named after the Welsh investor in John Cabot's expedition of 1498, Richard Ap Meurig, his name anglicised to Ameryke in the merchants' rolls.

Prince Madog's descendants were rediscovered in the nineteenth century: 'A stranger in the Mandan village is first struck with the different shades of complexion, and various colors of hair which he sees in a crowd about him and is almost disposed to exclaim that "these are not Indians!"' The tribe even gets into *The Descent of Man* in its account of the fit between the tint and the texture of the hair.

The Welsh astrologer and statesman John Dee used the legend to argue that Elizabeth I, as successor to the Cymric dynasty, had the right to the Americas. He laid claim to 'Sondrye foreyne Regions, discovered, inhabited, and partlie Conquered by the Subjects of this Brytish Monarchie' and identified Prince Madog as the first colonist of 'ancient Atlantis, no longer – nowe named America'. His attempt to use his own ancestors as the key to the New World did not, alas, succeed. Conveniently for the myth, the Welsh-speaking Mandan were wiped out by smallpox in Victorian times.

Fleure measured the heads of three thousand Welshmen in an attempt to trace their affinities and, with luck, to hunt down their kin in the Americas and elsewhere. In James James he found the perfect Celtic skull, the epitome of the short, submissive and spiritual people who had once filled the British Isles.

His notion was misguided. Head shape is under the control of a host of genes, which – like almost all their fellows – become diluted as different groups intermarry. Any inborn elements of the Celtic cranium are, like those for hair colour or blood groups, so much mixed with others that Welsh history is smeared across a landscape far broader than the principality. For most genes, Britain – Europe, indeed – is rather dull. A few trends pass from south and east to north and west, but in general the people of the far uplands of the offshore isles are not distinctively Welsh, or even British, but much the same as the rest of the Continent. De Gaulle's 'Europe of nations' gains no support from DNA.

Except, that is, for its men. Their Y chromosomes are witness to a chequered past, insulated from the rest of the population. Their splendid isolation has kept them pure. The Y is an arrow of manhood that flies from Adam to every male alive today. As mutations build up on its passage through history, each lineage gains an identity of its own. Fleure's mistake as he looked for the past in the bodies of the present was to forget that sex muddies the waters of descent. Safe from its influence, male chromosomes are a direct link to ancient times. To map them across the globe is to relive man's history.

James James was deficient in the name department, but surnames – as the family Sykes were the first to realise – illustrate how a masculine heritage can reconstruct the past. Among the Israelites, the Priests of the Temple took up the Cohen name soon after the death of Aaron, three thousand years ago, and in China such things were in wide use a millennium later, but in the West, surnames did not appear until the twelfth century, when a saint's name or an identity based on a profession or a birthplace was no longer enough to single out an individual.

In most places, surnames are inherited down the paternal line (the Chinese were strict, as women were allowed no names at all). Like genes, they change with time. James, itself the anglicised version of a biblical title, has mutated into Jamie, Games, Chames, Gems and many others. All these retain an obvious tie with their original, but a Mr Gunnison (a member of the clan and the founder of what is now a ski resort in Colorado) obtained his label from a misreading of that of his father, Mr Jameson. His own name made an immediate leap, far from its model. Just to be awkward, the Jacobis, a Jewish lineage, anglicised themselves to become James on arrival in the United States. They reached the same end as the others, but by a different route.

Not all surnames are as simple. 'Jones', for example, is a genealogist's nightmare. Family names did not catch on in Wales until the eighteenth century and, before then, children took the identity of their father. *Mab* means 'son of'. It mutates into *ap* or *ab*, to give labels such as ap Rhys (Price) or ab Huw (Pugh). Then the system was translated into English. Jones means 'Son of John'. Lots of people fitted such a broad description and many unrelated individuals (some not Welsh at all) found themselves with the surname. As the language has no letter 'j', John is spelled Ieuan – which became Evans or, for those who preferred the native system, Bevan. An English diminutive for John was Jankin – whose son became known as Jenkins. Wales is a Tower of Babel, with the same name shared by unrelated people and thousands who share common blood blessed with different surnames.

The Y shows that the Joneses are not alone, for plenty of families have a shared name but a separate ancestry. Many of the Pomeroys believe themselves to be descended from Cornish grandees of the same name, but the genes prove them to be a mixed lot, with separate origins in several places in England and Wales. The United States has thousands of citizens called Muma, Mumma, Mummau, Mummah, Moomaw, Moomau and more. A check of the crucial chromosome shows that many do spring from the same German root, but that others are from Ireland and even from Estonia. For them, too, a name is an ambiguous hint about the past.

James James himself left no sons, and local pedigrees are so tangled that it is hard to know whether any copies of his Y were passed on by one of his relatives. I myself do not have it, because his cousin John James Morgan was my mother's (and not my father's) father, which breaks the link. My own chromosome – and, needless to say, my surname – came from somebody else.

Whatever their ambiguities, names can say more about the past than merely identifying some long-dead paternal ancestor. In England, for example, names are less localised and less stable over time than in France where, just a century ago, the majority were found within a single *département*. The British, they show, moved about their landscape more than did their French cousins. What is more, the incidence of marriage between people with the same surname – a good measure of inbreeding – is much lower on the English side of the Channel.

Genes, like names, are ancient systems of identity. They are variable and change as the generations succeed each other; but unlike the surname they can preserve history over thousands of years. As the Y chromosome alters by mutation it records the past, on two distinct scales. It is well endowed with nonfunctional DNA. Most – like much of the rest of the genome – has a low mutation rate. Once in each few thousand generations, a DNA letter changes its nature, or a mobile piece of genetic material shifts to a new home. These large but rare mutations – 'Gunnisons' in the surname context – divide the chromosomes of the globe into a series of ancient and discrete classes. Hundreds of these haplogroups, as they are known, have been mapped across the world.

Male chromosomes are also filled with short sequences of DNA, repeated end to end, two, three or dozens of times. Such elements have a high rate of mutation, with an average shift of the number of repeats of about one in every three hundred new sperm. As a result, large differences build up in just a few generations. These haplotypes – the Gems and Jamesons of the genetical world – subdivide each haplogroup into a vast number of smaller lineages, which alter at some speed.

Y-linked haplogroups and haplotypes reconstruct the past far better than most genes, for – unlike variants scattered across the whole genome – the information they contain is not reshuffled each generation. As a result, the chromosome gives strong hints about the order in which mutations took place. The surnames James, Gems, Gemson and Gimson can be placed in only one logical series, and the same is true of the arrangement of repeated sequences on the Y. To work out the pattern of divergence in a sample of such chromosomes hence makes a much-branched tree of paternal descent.

Many people try to use their name to rebuild history. They spend years with parish records or passenger lists in the pursuit of ancestors and, for some, the hunt becomes an obsession. The fifty million who logged on to the 1901 UK census on the day it went online a century after publication caused the site to crash, and genealogy now gets more Web hits than any other topic except pornography. Genetics gives new fuel for its flames. Oxford Ancestors (with their 'Y-Line' test, courtesy of Dr Sykes of that noble university), Relative Genetics, Family Tree DNA and the others who have jumped on the band-wagon of history are all happy to oblige. At two hundred dollars and more a sample the search is not cheap, but a few days with the DNA can reveal more than years in the archives.

Surnames are clues about patrilines, in which husbands alone pass on a heritage (although for names at least, wives can join). Such a rule of descent was at the foundation of the Roman Empire and many other societies, but it contains a fatal flaw which ensures that any lineage, political or genetic, based on asexual inheritance cannot last.

A wedding brings together two people and four copies of most chromosomes. The marriage, though, has but one man and his Y. Male chromosomes are, as a result, only a quarter as abundant as are others. As their population is so small (and gets smaller if just a few of those who bear them are winners in the erotic battle) the risk of accidental disappearance of any male lineage, imperial or not, goes up.

Records of thousands of British families over the past five centuries show that, on average, a third of the men left no grandsons – which marked the end of their patriline and the loss of its chromosome. In small populations such random losses have a large and long-term effect. As one type disappears, others at once become more common and, as the process goes on, fewer and fewer patrilines are left. In time, accidental evolution of this kind causes isolated groups to diverge.

The search for sons among the Roman aristocracy (and in today's Japanese royal house, so far blessed only with a dynastically irrelevant daughter) shows the alarming readiness of patrilines, however powerful, to die out. The effect is large: not a single title of the five thousand feudal knights honoured in the Domesday Book survives today. The only way to rescue a badge of nobility, a surname or a gene is to give sex a chance and to allow the line to pass, now and again, through female hands. The Norman aristocracy (unlike their successors on the throne of England) were strict about the regulations, and William the Conqueror's knights soon faded from history.

Man's special chromosome shares the same fatal instability. As a result, male chromosomes, like grand families, are in constant battle against annihilation. The smaller the population, the more unstable matters become. A sudden drop in numbers leads to a permanent shift and, even if in later generations their descendants become abundant, they stay deficient in both the name and the gene department.

The *SRY*'s of the knights of Domesday may have fallen into statistical oblivion, but a small sample of those of their inferiors has – for reasons just as arbitrary – done the opposite. Some European Y chromosomes make up a far larger global share today than they did in 1492 because a small number of emigrants exploded in abundance as they filled the world. How abundant a haplotype – or a surname – might become in its new home reflects where its bearers came from, how many there were, and their marital success. New York has more names (and more chromosomes) from more places, in relation to its size, than anywhere else on earth while Duluth, in Minnesota, has far fewer. Both places are filled with migrants, but New York drew its men from across the globe, while the icy plains around the Great Lakes were populated in the main by Scandinavians.

The Y can say more about the past than the noblest of titles but is not always an unambiguous guide through the years. Some parts change fast and blur the record, and to confuse the tale further, the same genetical end can (as for the Jacobis who turned into Jameses) be reached through different means. The Darwinian rules also wipe clean the slate of history. With no opposite number to protect them, all Ys are at risk of disaster when exposed to selection's purifying flame. In Japan, males of a certain haplogroup have low sperm counts and often turn up at infertility clinics. Their chromosomes, abundant as they are today, may soon disappear and their failure will confuse future attempts to reconstruct the nation's history. The abundance of any haplotype hence reflects an unknown record of natural selection as well as a slow build-up of mutations balanced by random loss. The Ys of the world are so similar that some biologists believe in a universal Adam who may have lived just sixty thousand years ago. The hint is vague at best.

Over shorter periods the chromosome is a more reliable clue. In some places it confirms more traditional records of the past. Thailand retains two distinct patterns of marriage. In some hill tribes – such as the Karen, a people displaced from their main homeland in Burma – women stay in their home village and expect their husbands to join them. In other groups such as the Akha (who migrated long ago from Tibet) sons stay at home to await the arrival of a wife. For the Karen, mitochondrial genes – inherited through females – differ from village to village while the genes for male identity do not vary much across the map. For the Akha the converse is true.

The genetic geography of Thailand confirms the deep roots of its modern social patterns. DNA can also reveal the marital habits of peoples who died out before history began.

Plenty of European men, encouraged by bestsellers about self-discovery, see themselves as fearless explorers *manqués*. They regret the loss of a supposed past in which their ancestors ranged unfettered across a great landscape, to spread their seed to a series of passive and expectant mates. Alas, the molecules prove the idea to be a myth. The European map of the Y looks much more like that of the Akha than of the Karen, with three times more geographical variation in male genes compared to those of their partners. Adventurous as its men may have been over the last few colonial centuries, for much of the past they stayed at home. 'Husband' comes from the Old English 'house dweller', and – in stark contrast to their image as rapists and pillagers – for most of the time, that is where men remained.

Husbands travelled less than wives for reasons of fiscal prudence. Farmers pass their land to their sons. Their wives come from near by to join the household and, over the years, genes move further through females than through their partners.

Man's ancient lethargy lasted almost to the present day. Italians still follow the habits of the Roman Empire. Nine-tenths of bachelors live with their parents (or in the same block of flats) and in many places a quarter of married couples live in the husband's childhood home, for, as they say, 'A woman's house is a prison'. In their reluctance to move, Italian males follow the French philosopher Blaise Pascal. A man's unhappiness, he said, begins when he leaves home. All over the world, men have taken his advice. For thousands of years, three out of four societies followed the Roman pattern and daughters moved to join a patriline while sons stayed where they were born.

Then the economic balance changed and the European Y went on the move. The dismal science forced millions of its bearers to leave home. As poverty drove Italians and many others from their native lands, names and genes poured into the New World. The sons of the Roman Empire made their mark in New York, as any visitor to Little Italy can see, and in the south as well. Even today, the masculine heritage of parts of Brazil has, in terms of names and genes, strong affinities with the descendants of Romulus and Remus. The British were just as bold. There are plenty of Sykeses in the Americas, even more people called James, and a plague of the family Jones.

Millions of Americans have ancestors in Europe, and the genealogy industry turns on their hunger to find the details of their past. Once, most of the links were made with Ireland, Sweden and Germany, the mother countries of the majority of those who fled across the Atlantic. Nowadays more and more of the nation's ties emerge from further south. In the 2001 United States census, one resident in eight identified himself as of Hispanic descent, a proportion far higher than even twenty years earlier. Their surnames and their native tongue may reflect a Spanish root, but most Hispanic

Americans can also trace a good part of their ancestry not to the conquistadors, but to a far earlier group of travellers.

In October 1492 Christopher Columbus made a landfall on the island of San Salvador in the Caribbean. He at once recognised it as the Garden of Eden. Few US citizens still see their land as a pre-lapsarian paradise, but Native Americans rejoice in a link with their continent's soil that began long before Genesis. The extensive exchange of genes between the indigenous populations of South America and their Iberian conquerors means that many of the Hispanics who have moved to the United States could make the same claim. Their new abundance, and their Native American genes, has brought the nation closer to its pre-Columbian roots than it has been for many years.

The ancestors of the Americas were once assumed to spring, like those of its later migrants, from across the Atlantic, from the lands of the Garden of Eden. All kinds of people made the trip. Prince Madog, needless to say; but Vikings, Basques, Arabs and more were thought to have floated across.

In 1996 a remarkable skeleton was found on the shores of the Columbia River in the state of Washington. It boosted the notion of an ancient transatlantic journey. Kennewick Man appeared to have European features, with a long head and a prominent nose. So different was his skull from those of American Indians that it was assumed to belong to a nineteenth-century pioneer. Judicious use of plasticine by the intellectual descendants of H. J. Fleure gave Kennewick a face – and he had the features of an American entrepreneur. Life had been hard. His skeleton was battle-scarred, with fractured ribs, a withered arm and an injured hip. The fatal wound was caused by a leaf-shaped stone spear (a common problem among the Americans of those days, with warfare the cause of a third of adult deaths). Radiocarbon dating proved the bones to be more than nine thousand years old.

At once they were hailed as evidence for an ancient European link. Perhaps the myth of Welsh ancestry in the

Americas was not as absurd as it seemed. Quite where in the Old World Mr Kennewick may have come from could not be determined from his skeleton, which yielded no DNA. His remains were claimed by various supposed descendants, including the Asatru Folk Assembly (a council of American Vikings who have revived the ancient Nordic religion). They went so far as to conduct an Odin–Blot ceremony over them (with apple juice rather than mead as alcohol was not allowed in the local museum). White supremacists hailed the discovery. The United States was, they claimed, a nation founded by Europeans, and the Indians should cease their complaints about the theft of their land.

Attempts to look further into Kennewick's supposed Viking (or even Welsh) affinities have been frustrated. To discourage visitors, the US Army Corps of Engineers hid the burial site under five hundred tons of gravel. The bones themselves were for a time deemed to be subject to the Native American Graves Protection and Repatriation Act, and were returned to the Nez Perce and Yakama tribes, who claimed 'cultural continuity' with the Washingtonians of nine millennia ago. Although a Federal Court has reversed that judgement, no anthropologist has yet had a chance to examine them.

Whatever the truth about Kennewick, fossils show that the New World was colonised long before his day, about sixteen thousand years ago. Those ancient travellers came not from Europe but from the other side of the globe.

From Siberia, they traversed the Bering Straits (then dry) to reach Alaska, and, within a mere two thousand years, filled their new-found land to its southern tip at Cape Horn. Their remains are scarce, and their stone tools almost as hard to find. The bones are surrounded by wrangles about dates and some of the supposed tools by doubts about whether they are artefacts or just rubble. Now the genes have told the truth about Kennewick. They forge an unexpected link between the two shores of the Atlantic Ocean.

Y chromosomes reconstruct the American journey in all its stages. They show how perilous the trip across the icy isthmus to Alaska must have been. The same haplotype – the legacy of a single individual – is shared by half of all native North American males as a hint of how few travellers struggled across the Bering bridge. The voyage onwards was also hard, with lots of bottlenecks on the way. Four out of five South American Indians carry the same version of the Y, proof that even smaller numbers managed to complete the traverse across the Equator. In the North (but not the South) some genetic surnames are so distinct as to suggest a second and much later Asian migration.

Genetics tells a tale which, in general, confirms the archaeological record of the people who moved from Siberia to Alaska, and on to the furthest tip of Patagonia. Genes, though, say much more than anything hinted at by the fossils.

Some of America's patrilines are still found in Asia. One is common, as might be expected, in the Siberians who live close to the Bering Straits. Now the roots of the American male can be traced far deeper into the soil of the Old World.

The Kets are a people of the Yenisei river basin in the Krasnoyarsk province of eastern Russia, in the heart of Asia, three thousand miles from the Bering Straits. Two centuries ago they were abundant; a nineteenth-century traveller described them as plump with thin legs and a staggering walk, flitting eyes and a jerky talk. They looked, he said, rather like Finns. Today a mere thousand or so are left. Most live in poverty and are in thrall to the vodka provided by traders in return for furs. Their language is on the edge of extinction, with a few hundred speakers and a desperate attempt to keep it alive. It is the last member of a once prosperous group – Yug, Pompokol, Arin and others – which, even a century ago, were spoken over much of southern Siberia. They were quite unlike any other in the region, and their verbs had up to eighteen separate sounds. Ket is a distant member of a vast confederacy of tongues, which includes Chinese, several Caucasian dialects, certain languages of the New World – and perhaps even Basque.

'Ket', in Ket, means 'man'; and Ket men, their chromosomes show, have a close tie with the natives of the Americas. The commonest haplotype in the New World – borne by the majority of Native American males – traces straight to the Kets and is absent from the tribes who surround them. The aboriginal Americans hence find at least some of their forefathers in the heart of what was until not long ago the Soviet Union.

The spokesman for the Confederation of Tribes of the Umatilla Indian Reservation did not need biology to find his own roots. He justified their actions to stop research on Kennewick with a proud claim: 'Our sacred human remains should not be a product to generate data. We already know about our past. We have oral histories that go back ten thousand years. We know where our people lived, how they lived, what they lived by.' Science, it seems, is beside the point.

The Welsh, alas, are less confident about their legends. Their identity is wrapped up in a language whose prospects are almost as uncertain as are those of Ket, and their memories of a royal voyage to the land of Kennewick – if it were ever made – are incomplete at best. Madog and the Welsh tribe in the Americas have each sunk into myth.

Whatever the truth about ancient Welsh explorers, millions of their kin made a later journey to the west. A great mass of Celtic genes crossed the Atlantic in the nineteenth century's waves of migration.

Philadelphia translates its name as the 'the city of brotherly love'. Its territory, Pennsylvania, its citizens believe, gained its own title from a concoction of the surname of the seventeenth-century Quaker William Penn and the Latin word for forest. They are wrong, for the Liberty Bell State has Welsh roots.

Penn's colony attracted many migrants, among them a

group of Welsh Quakers who hoped to set up a 'barony' in which they could speak their native language, under threat even three centuries ago. Penn set aside land, the 'Welsh Tract', for his fellow countrymen. Its thirty thousand acres were divided among Messrs Jones, Lloyd, Evans, ap John, Wynn, ap Thomas and Davies. William Penn then – to their immense annoyance – reneged on his promise and the Welsh-speaking colony never materialised. Even so, the language itself was often heard in Philadelphia a hundred years later.

When it came to a title for the territory as a whole, Penn suggested 'New Wales'. The emigrants disagreed (a New England, they felt, was quite enough) but Penn appeared to compromise. The colony would be named in part in Welsh: 'Penn being Welsh for head as in Penmanmoire, in Wales . . . I called this Pennsylvania, which is the high or head wood-lands' (the extra 'n' was no doubt a clerical error). Today's suburbs of Philadelphia include places called Uwchlyn, Llanerch and Tredyffryn – names that come not, as often assumed, from Native American tongues, but from Wales.

Welsh influence in the New World goes well beyond the suburbs. Sixteen of the signatories of the Declaration of Independence were of Welsh origin (as Thomas Jefferson himself wrote, 'The tradition in my father's family was that their ancestor came to this country from Wales, and from near the mountain of Snowdon, the highest in Great Britain'). Other prominent Americans can claim a similar ancestry (Presidents Monroe, Lincoln, Coolidge and Nixon and the comedian Bob Hope among them). Over the years, the inhab-itants of what never became New Wales were overwhelmed by later migrants and faded from view (although some of their descendants still search, perhaps rather forlornly, in the records to find out from which particular Jones or Evans they might trace themselves).

Other Welsh Americans retain a stronger link across the seas. Their ancestors migrated to the far end of the continent, to Patagonia, and their descendants have retained much more of their identity than have their cousins to the north.

In nineteenth-century Wales it was a punishable offence to speak the language in school, and the nation felt its whole culture to be under threat. Michael Daniel Jones of Bala had taken a group to join his fellows in the United States, but even there they were swamped by anglophones. He returned home and encouraged a new move to Argentina, to the unconquered regions of the far south. In 1865 the first shipload of a hundred and fifty emigrants, most of them men, sailed aboard the *Mimosa* from Liverpool.

After a hard time in the salty marshes of Puerto Madryn, they moved inland to the Chubut Valley, with a second settlement at Cwm Hyfryd in the foothills of the Andes. More settlers soon arrived. They built the first railways and watered the desert to grow wheat. Attitudes had changed since Darwin's visit to Tierra del Fuego, five hundred miles to the south, thirty years before ('I could not have believed how wide was the difference between savage and civilized man; it is greater than between a wild and domesticated animal . . . Viewing such men, one can hardly make oneself believe that they are fellow-creatures'). The Welsh got on well with the natives, and the contempt and slaughter the Patagonians had suffered at the hands of earlier invaders was succeeded by trade, and in time by intermarriage.

The presence of Welsh patrilines on the edge of the South Atlantic today is far more noticeable than in William Penn's colony. There are hundreds of Joneses and Jameses in the local telephone books; names which, without doubt, are accompanied by Welsh chromosomes. Their bearers try hard to maintain elements of Celtic culture (bardic gatherings included), with plenty of speakers of the language and much tourist interest in Welsh chapels and tea shops. Their gift shops are full of Celtic souvenirs, which include silver buckles and Christian crosses decorated with fine whorls, spirals and lines.

The originals were once traded across much of Europe and replicas are everywhere in Wales itself. But who first made them, and what, if any, is their tie with the Welsh of today?

Celts are, in several ways, an ambiguous group. *Keltoi* was a term used by Greek historians to describe a variety of savage tribes who lived along the Danube. The Romans picked up the word, but their *Celtae* were inhabitants of southern France, who had the unpleasant habit of divining the future from their enemies' struggles as they were sacrificed (and of taking their heads home as souvenirs). Later the noun became a derogatory term for barbarians in general, and then more or less disappeared. Not until the seventeenth century, when it was realised that Welsh, Erse and Scots-Gaelic all descend from a Gaulish tongue once spoken in France, was the term revived, this time to describe an extinct language. Inevitably, it was soon used to suggest a common heritage among those who speak such dialects today, Welsh included; in modern terms, to infer that the Celts share a set of genes that distinguishes them from their neighbours.

Dwr is the Welsh for water – and the failed medieval liberator of Wales, Owain Glyn Dwr, called himself after a brook of that name. Other rivers hint at the breadth of his ancient empire. Sussex has the Adur, France the Adour, Germany the Oder, Hungary the Tur, the Czech Republic the Turie, and even Zurich might have gained its title from the same root. Excavations near Ankara in Turkey, at ancient Gordion (the home of the Gordian knot), reveal the presence of Celts there, too. By the third century before Christ the Galatians, as they were known – a people much written to by St Paul – were in residence. They came as mercenaries, but stayed on. Their two-faced statues and traces of human sacrifice tie them to their European kin. The Celtic language was, the relics show, well established from Turkey to Wales, where the great barrows and hill forts built by those who spoke it can still be seen. Within them are buried the intricate swords and shields of the ancient Celts themselves. Such artefacts, and Celtic culture in general, reached their height at La Tène on the shores of Lake Neuchâtel but, for a time, filled an area almost as large as modern Europe.

The imperial ambitions of the Romans pushed their way of life to the margins – which, in Britain, meant the mountains of the north and west. Slowly, Christianity took hold, for the western Celts at least. Now crosses rather than weapons were decorated with spirals and geometric patterns. After the final decline and fall of the Roman Empire, a second wave of invaders, the Anglo-Saxons, came to Britain. By AD 600 the fragile political unity of the natives was crushed and they fell back before the new culture.

The ancient Britons spoke with much bitterness about their plight in the face of the barbarians: 'The germ of iniquity and the root of contention planted their poison amongst us, as we deserved, and shot forth into leaves and branches. The fire of vengeance spread from sea to sea, fed by the hands of our foes in the east, and dipped its red and savage tongue in the western ocean.' Thus wrote the sixth-century Celtic monk Gildas Bandonicus, not an admirer of the Saxons or of their cruel and irreligious way of life.

Cymry (the Welsh term for their nation; 'Welsh' itself derives from the Saxon word for stranger) made its first appearance in his time as a hint that the Britons had at last begun to identify themselves as a single unit rather than a set of warring clans. But what did that mean? Was their identity based on a shared mode of speech (as English today unites Britain and Nigeria) or on a deeper tie of descent? The Celts have been defined by barbarism, by geography, by artistic style and by language. Any biological links among them, or with the modern Welsh, were, until the new genetics, quite obscure.

The history of ancient Britain as it went through colonisation and a shift of language was repeated again and again as Europeans filled the world, in South America as much as anywhere else. The Welsh journey to Patagonia seems at first sight just a minor detail in the record of expansion. Y chromosomes tell a different and a greater tale. They reveal the true affinities of Gildas Bandonicus and his fellows and uncover an ancient connection between the people of Wales and the natives of the Americas, in a story of genes that circle the world to tie together figures as disparate as Prince Madog, Kennewick Man and James James himself.

Geneticists recognise several hundred distinct Y haplogroups, the large families of variants defined by ancient single-gene changes. One of the first to be found – haplogroup 1 – varies greatly in abundance from place to place. The molecular surname is quite common in England but becomes much rarer south and east across Europe. In Syria (where I myself once persuaded, with some help from the powers that be, a number of men to spit into tubes) just one in a hundred bears it.

In Wales, in stark contrast, nine out of ten men carry some version of haplogroup 1. A single subtype is responsible for 70 per cent of Welsh male chromosomes (which gives the nation the most homogeneous masculine identity in Europe). The variant is also frequent in Ireland and among Scots with Celtic rather than English or Nordic names (although Scotland has a large input of Viking and Germanic varieties), while Cornishmen, in spite of their sporadic nationalism, are impossible to tell apart from their English rivals. The shift at the English border is abrupt and over a few miles the frequency of the Welsh haplotype rises by half. James James and his kin – whether born on the slopes of Plymlymon or of the Andes – almost certainly bore the shared and ancient label of their Celtic forefathers.

The narrative of the Welsh Y (like that of the Native Americans) fits, up to a point, the standard view of the past. The ancestors of today's Welsh once filled the British Isles, but faced an influx from the east by Anglo-Saxons and others who pushed them to the margins and to an embittered contemplation of their glorious days long gone.

Once again the genes say more than anything in the records. First, they reveal a great difference in the history of the sexes. The Ys of the Celtic fringe are distinct from those of the rest of the British Isles, but their mitochondrial lineages (inherited through females) and the genes passed through both parents are almost identical to those of England. Most of the DNA of the native Britons, it seems, was absorbed by the invader.

However, half the population stayed aloof. The record of the Y shows how the Celtic libido prevailed in the face of political change. The small dark men of the west kept the Anglo-Saxon male chromosomes (and those who bore them) at bay, but welcomed their wives and daughters. Welshmen, says the DNA, were able to retain their sexual identity in the face of a wave of alien females who rolled over them.

Haplogroup 1 sets Wales and Ireland apart from the later invaders of these islands. That may gratify nationalists of various flavours, but DNA ties the men of those places to some distant and unexpected relatives.

The Basques – who, like certain Celts, try to endorse their national integrity with dynamite – have long been known to be distinct from their neighbours. Their genes stand apart, and their language too (unlike Welsh, which belongs with Sanskrit, Greek and English in the Indo-European family) is in a group of its own. All this supports the Basques' view of themselves as a remnant of ancient Iberian hunter-gatherers who held out against the farmers as they moved in from the east, long before the Celts themselves emerged. They descend from a culture so old that almost no remnant is left. The names of the rivers Isar in Germany, Isère in France, Yser in Belgium, Jizera in the Czech Republic and Esera in northern Spain relate, in all probability, to water. They resemble no word in any extant tongue, and are remnants of a language destroyed by the farmers and their Indo-European speech. The Basques refused to treat – or speak – with the incomers and, as a result, stayed separate for five thousand years. Their disdain was more or less absolute and most of their genes are still distinct from those of the peoples around them.

The Basques are not as unique as they think, for they have relatives in the north. Their male chromosomes are almost identical to those of Wales. Welshmen, it seems, share masculine kinship not with the descendants of the Celtic cultures of central and southern Europe who gave them their language, but with an older and fiercer people now confined to northwest Spain.

The tale written in Welsh DNA reveals the ambivalent attitude of the British aboriginals to a new way of life that arrived around five thousand years ago. The locals may have kept their Y chromosomes to themselves, but they soon picked up the alien economy (and its women). They abandoned their own speech (which might have sounded rather like Basque) in favour of that of the farmers. Welsh itself is no more than a dialect of Indo-European which was driven to the edge by a close linguistic cousin, English. The Celtic tongue has become the icon of Welshness (with a promise to preserve it included in the national anthem) but it is a newcomer.

The patrilines of today's Welsh and Basques are the direct descendants of those of the earliest Europeans of modern form. Those pioneers spread into a deserted continent in a mild period between two ice ages, some fifty thousand years ago. Modern European culture began about then, with the Aurignacians. They left chipped blades and scrapers of flint, which were soon supplemented with decorated tools made from antlers, teeth and the like. Thousands of miles to the east, another civilisation was on the move. Siberia was warmer than today, with forests of oak and elm rather than pines. The first humans of modern form, with their simple stone tools, appeared in the southern part of its great landscape at about the same time as the Aurignacians. Soon they, too, came up with art, based on ochre, bone and ivory ornaments.

For a time the Aurignacians and their Siberian equivalents hunted deer and gathered berries, but then the glaciers returned for the last time. The bitter climate allowed people a mere toehold (albeit one interrupted by flurries of cave painting). It drove them from northern Europe altogether, but much of the Siberian plain became an icy 'mammoth steppe' where men hunted elephants and other mammals. At the peak of the great freeze, around eighteen thousand years ago, even those hardy Asiatics were driven back from what had become a chill and windblown desert.

They found a refuge around Lake Baikal, close to the home-land of today's Kets. Over hundreds of thousands of years, its vast bulk of fresh water has never frozen throughout the year. It has been a safe haven for life of many kinds. As the world warmed after the last glacial advance, the pioneers spread from their sanctuary. Some chased the ice as it retreated north. They reached the hostile lands of Siberia and moved onwards to the Americas and, in time, filled the New World to its tip.

Kennewick himself probably carried one of the Y types now present among Native Americans, the most frequent of which traces back to the Kets and their neighbours. Central Asia is still a centre of diversity for Y chromosomes, and the oldest version of the famous haplogroup 1 is found there. Several variants separated from it by just a few steps are common in the natives of the New World. The Asian genetic signature links its bearers not only with the first Americans but, quite unexpectedly, with the inhabitants of the western fringe of Europe.

The accumulation of mutations on the Y can be used as a (rather inaccurate) clock to date when its various lineages began to diverge. The Welsh and Basque haplogroup 1 chromo-somes differ at a small number of sites from a type abundant in the Ukraine and in turn are just a few steps removed from the Y chromosomes of the Kets. The molecular clock dates the origin of that ancient line to well before the last ice age, forty to fifty thousand years ago, and ties the people of ancient Siberia to their European peers.

The ancestors of the Kets were, male DNA proves, a source not just of the American Indians, but of the first post-glacial hunters of Europe and, as a result, of the settlers of the British Isles. Native Americans are, it shows, linked with the people of Wales through their Y chromosomes. They make a genetic connection between Kennewick Man and Prince Madog – not drawn across the Atlantic, but across twelve thousand miles of Europe, Asia and Alaska. Madog's genes are in the New World, but got there via a path quite different from that imagined by legend.

Fleure himself was pessimistic about the survival of his mythic entity: 'The Cader Idris and Plynlymon ranges long provided a barrier, which is likely to count for less and less in these days of powerful charabancs, listening-in, and universal education.' Those words were written fifty years ago; and the charabancs have moved on. They bear Welsh genes across the world and their Saxon equivalents into Wales. Even so, the Welsh (or at least the males of the tribe) have managed to retain a certain genetic individuality into the third millennium.

They can, through male molecules, trace themselves as native Britons back to the days when the land last emerged from the ice. Their past is shared with a remote people who provided the founders of the New World. Their descendants in the far south, when they welcomed the emigrants on the good ship *Mimosa*, were as a result (and quite unaware) taking part in a family reunion. The circle closed with the marriages between Welsh men and the women of South America.

Fleure would be delighted to learn as much. I have checked my own Y, and it too reveals an origin in his beloved tribe, the Little Dark People. I was, of course, happy to find a direct link through the Celtic mists from my own puny frame and once black hair to the aboriginals of these isles, to their Patagonian descendants and, through a more tortuous route, to the natives of that southern land.

Fleure's stereotype has some truth, but I tend to steer clear of religious and bardic gatherings – and, unlike James James, received not a penny for my small contribution to Celtic history.

CHAPTER 9

POLYMORPHOUS PERVERSITY

Sperm can be expensive stuff. Americans who hope for artificial insemination are obliged to pay hundreds of dollars for a sample – and more for a high-quality source. The Repository for Germinal Choice, which specialised in Nobel Prize-winners, collapsed after two hundred births, but other spermatic academies remain open. When it comes to the vital fluid, the market has spoken; a shortage means high prices. Even countries with a free system of blood donation offer cash to those who give sperm. The sum is often modest, at around fifteen pounds a shot in Britain, but essential if the precious fluid is to be found. Two-thirds of British benefactors do it for money alone, with no idea of altruism. Indeed, a typical donor would be happier to ejaculate for free if his efforts are to be used for research rather than to produce a child.

And that is odd, for the basic tenet of sociobiology – the attempt to understand our own behaviour with reference to the animal world – is that men, like beasts, try to increase their fitness by spreading genes as far as they can. They are constrained in their promiscuity only by the expense of raising a child, the vehicle of their DNA. But a sperm donor need not worry, for he has all (or most) of the joys of insemination while somebody else picks up the tab. Where are the queues of men with test tubes, each eager to increase his genetic input to the future, for free?

We are animals; and in genetic terms can scarcely be told from chimpanzees, mice or – give or take a few DNA bases – worms. Fruit-fly genes shed light on inherited disease, and yeast does the same for cancer, so why not use flies, or yeast, or gorillas, to understand our reproductive condition, if not today, at least in some primal state? Several chapters of *The Descent of Man* are devoted to the evolution of what Darwin calls mental powers and moral faculties, and the book describes numerous instances of what he saw as both noble and ignoble behaviour among apes. W. S. Gilbert's Lady Psyche is at one with the idea: 'Man will swear and Man will storm/Man is not at all good form/Is of no kind of use/Man's a donkey/Man's a goose.' Geese and donkeys are not often used as role models, but the habits of many birds and mammals seem to echo our own. Perhaps the key to man's behaviour can be found by placing him in context.

Animals can illuminate the lives of men, but the problem is to know when to stop – when to recognise when hypothesis becomes speculation (or, worse, advocacy). Too many attempts to explain behaviour in evolutionary terms depend on untestable events long ago. Freud claimed that children go through a stage of polymorphous perversity, in which pleasure shifts from mouth to anus to genitals. Only in the last phase does the infant become mature. A failure to achieve it damages the inner mind. To analyse a forgotten childhood can, as a result, be the key to an adult's neuroses.

If human nature once flourished unbound, the natural world might tell us what, beneath the veneer of civilisation, we are really like. The argument, like that of Freud, suffers from the infinite diversity of the past. To rake over all incidents from childhood is bound to turn up experiences that can be used to explain any later event (the links need not be close; Little Hans's desire to sleep with his mother came from his distant memory of a giraffe). In the same way, the polymorphous perversity of life offers a fatal temptation: to explain ourselves with an anecdote that fits.

Animals help us to understand how sex began and how its machinery works. They illuminate the science of maleness, but men need something more. We are less constrained by ancestry than is any other creature and can (as California in the 1960s and Afghanistan today each show) transcend the biological rules by choosing what we should do, with the use of criteria denied to all other beings. Sociobiologists tend to forget that tiresome fact, under the impression that they are doing science.

Although the natural world is not a universal excuse, it remains a source of innocent amusement. It can go further, for when the actions of animals are put into context they may hint at the fundamental rules of sexual behaviour.

That pastime has long been a source of shaggy dog stories (some of which even involve dogs). Natural history consists of little else. Darwin himself was fond of its excesses. The males of some barnacles were reduced to hitchhikers inside their mates: 'The female . . . had two little pockets, in each of which she kept a little husband; I do not know of any other case where a female invariably has two husbands.' Some contained fourteen of those happy creatures at a time.

To the untutored eye – and to non-Darwinians – such endless coital tales seem little more than a genetical Ripley's 'Believe It or Not'. A turkey makes a droplet of semen just visible to the eye, while Grevy's zebra delivers half a gallon. The fire ant pumps out enough cells on each attempt to allow his mate to produce two million young over seven years while in a certain fruit fly size counts, and each sperm – as long as a finger – is rolled up and handed over one a time. Hedge sparrows take a tenth of a second to copulate, but a shiny brown desert beetle found in Australia spends weeks at the business (and is happy to pass the time in embrace with an empty beer bottle).

The penis, too, is infinitely flexible. In absolute terms, the blue whale wins with a ten-foot organ. Darwin was always a little coy about the details in books aimed, like *The Descent of Man*, at a general audience; but in more technical works, such as his vast tome on barnacles, he spoke with much admiration of the 'wonderfully developed' member of certain kinds, which was several times their own body length. Snakes have two penises, used in turn, but other animals cope without the organ. (Newts, for example, make a bag of sperm, which their partners pick up if so minded). Squid use a specialised arm as an explosive sperm–delivery system and Japanese diners have hurt their mouths when they eat one cocked ready to fire.

Such stories are endless, but Darwin turned them into science. His notion of sexual selection depends on the relative outlay of each parent in the offspring. To gain a mate, males have high fixed costs, be they antlers or sports cars, but after the initial outlay are not obliged to contribute much more. Females, in contrast, face a heavy demand for investment in each child as it appears. In stockmarket terms, males are bulls and their partners are bears. A donor of sperm must put in a lot to have any hope of success, but above that minimum a small increase in outlay offers dramatic returns. For him, a gamble on future success is always worthwhile. His mate is more cautious: she may complete the course on each attempt, but the expense is so high that each try calls for a safer bet.

Selection works hardest on the parent that does least to raise the young, because it must gain access to the capital put in to the task by the other party. That feckless individual (often, but not always, the male) faces competition in the erotic market place from its fellows who are anxious to strike an even cheaper bargain. The vendors of sperm must, like unsavoury stockbrokers, invest in pricey displays to attract a customer or to drive the opposition to bankruptcy. Males take risks and females make choices, and the stag's antlers, the mandrill's bottom and other beauties of life arise from the balance between one party's urge to promote its genetical assets and a certain caution by the other about its quality.

Animals make their sexual deals in ingenious ways. The bluegill sunfish, an inhabitant of American lakes, lives on deception. Some males are large, aggressive, and hold a territory that contains several mates. They fan the eggs to provide oxygen and scare off predators when the young hatch. Those noble fish face tricksters: either cheats who rush in whenever they get the chance, or transvestites who look like females and cruise unnoticed through the master's patch, scattering sperm as they go. DNA shows that the large animals are often duped and that it pays as well to be furtive or to cross-dress as to be a paterfamilias. Don Juans do no better than Machiavellis in the reproductive struggle.

Until not long ago, the common view of the makers of sperm, be they fierce or sly, was as masters who pressed their attentions on partners who had rather little choice in the matter. The index to *The Descent of Man and Selection in Relation to Sex* makes the point well, with its 'On the pugnacity of the peacock', 'Birds, eagerness of the male in pursuit of females', 'On strife for women among the North American Indians' and many more. Darwin accepted female choice, but was impressed by male vigour. Now it seems more and more that the consumers drive the market. Sperm donors may be crafty, but its recipients are cautious. Females have more control than once appeared and can often force their reluctant mates along an arbitrary and unwelcome path in the search for approval.

Males invest their resources in various ways, some obvious, but most less so. At two and a half tons, a male elephant seal weighs seven times as much as his partner (which is equivalent to a woman of normal size having an affair with the world's fattest man, Walter Hudson, who died at half a ton). Unlike the unfortunate and immobile Mr Hudson, elephant seal males are agile, fierce and for much of the time angry.

Just one in five reaches maturity and even among the survivors just a few pass on genes. Nine out of ten have no progeny at all, and three-quarters of the females are inseminated by a twentieth of the potential mates.

The elephant seal in his erotic fury is a monument to feminine, rather than masculine, power. Much as the larger animals might appear to be in charge as they inflate their noses, roar and batter each other until the blood flows, in fact their consorts force the best to choose themselves. At first, females reject anyone who tries to woo them. Any individual bold enough to try to copulate gets a firm brush-off which attracts the attention of others, who soon begin to fight. The object of attention stands coyly aside until the best seal emerges, bloody but unbowed, and allows him to mate. The battle was his, but the choice was in effect made by her.

Some conflicts are even bloodier than those of the elephant seal. In the duck-billed platypus the sex ratio shifts from around even at birth to six females to each male as older animals use poisonous spines to kill off their young rivals. Any bearer of a Y chromosome must balance desire against health, as somebody who has just been defeated by a two-and-a-half-ton rival (or a poisonous senior citizen) is rather less attractive than before.

Males may use more subtle tactics in their battles for success. Insects insert plugs and pessaries to block the access of a later hopeful (who may in return evolve hooks and pumps to solve the problem). Sometimes they are fooled, for the cells stick to the extraction pump itself, and fertilise the next animal to receive it. In an obscure group of worms, competitive corking goes further, for the top animals mate with their male inferiors and plug them to frustrate their hopes. Others, like Grevy's zebra, try to flood out a predecessor's donation. The right whale goes in for the same strategy, with testes that together weigh a ton and produce a huge (but as yet unmeasured) ejaculate. Certain butterflies are more ingenious and make vast numbers of sperm empty of DNA which stop the passage of those from a later encounter.

The sex act is but an overture to a genetical grand opera in which those who give birth set the rules, while their partners, belligerent as they might be in the first few bars, become more and more passive as the plot unfolds. Males are often forced to submit to the painful truth that, in the end, the consumer chooses.

Sometimes females weigh the value of a gift – a piece of food or a bright feather; or, as in a certain moth, a cloak filled with poisons able to keep spiders at bay. Sperm can itself be a useful bequest, and some female fruit flies are happy to soak it up to increase their output of eggs. Certain crickets go further, and wrap the donated cells in a tasty wrapper. The great French naturalist Henri Fabre wrote of the praying mantis that it made the ultimate sacrifice and was devoured by his spouse after copulation. As he put it, 'If the poor fellow is loved by his lady as the vivifier of her ovaries, he is also appreciated as a piece of highly flavoured game.' His prose was better than his biology, as mantises do not often go in for such behaviour – but plenty of snails, insects and spiders do, with the lethal Australian redback well to the fore.

Males are also judged by their ability to pay the taxes intrinsic to their state. The essence of masculinity exerts a levy of its own. Testosterone makes animals aggressive, but also suppresses the immune system. In starlings, a quick injection of the hormone cuts its efficiency by half. Reindeer, too, suffer most from the attacks of parasites at the time of the rut, when they are abuzz with the stuff. Sometimes, females can assess a mate's ability to cope with his own chemicals. The bright plumage of jungle fowl infected with roundworms fades as their enemies take hold. The immune system struggles to keep up – and the animals who lose most colour are at once rejected. Bright feathers are expensive, and so are antibodies, and only those with lots of capital can afford both. Females, as they choose those able to sustain fine decorations even when filled with parasites, make a test of the quality of the genes on offer.

In the battle of the sexes, plumes, blubber and self-immolation are visible enough, but the struggle goes on well after solid flesh has done its job. The efforts of the sea elephant and his fellows are as nothing to what takes place later. Once within their partners, with curtains lowered, masculine cells are forced to yield to feminine choice and the efforts of those who made them are revealed as mere icing on the procreative cake.

Darwin himself noted as much. In *The Descent of Man* he writes of his ecclesiastical cousin, William Darwin Fox, who kept geese of the common breed, together with a Chinese variety. A Chinese male pressed his attentions on one of the common birds; and, in the clutch of eighteen eggs, fathered fourteen of them. 'The Chinese gander', he wrote, 'seems to have prepotent charms over the common gander.'

Prepotency comes, in the main, not from ganders but from geese. Females go a long way to choose the sperm they prefer. Chickens eject much of the ejaculate at once so that most cells do not even start the race. Even when the first hurdle has been passed, the ardent travellers must swim through a tortuous internal tract, and are diverted into storage tubes, each with a filter at its mouth, where they can stay for days or weeks. Many of those who enter the haven are damaged – they have two tails, or a battered head – but all those released seem to be perfect. To store sperm allows a female to sample several mates before she decides who gets into the inner sanctum. In one experiment, the donations of fifteen male turkeys were mixed, and a bird inseminated with the cocktail. The recipient sifted out the second-rate cells, and those of a single animal fertilised the entire clutch. As a hint of the subtlety of choice, showy and dominant cock jungle fowl attract more mates, but their subordinates' sex cells do better once inside.

All this is in accord with Darwin's logic: with competition by one partner and choice by the other. Remarkable as such conflicts are – and there are thousands more – science is not science without theory. In biology, simple facts are of little use until they are put into context.

Evolution is, above all, the science of comparison. To reconstruct the past from the present is not easy, as the endless perversity of nature makes it easy to pick examples to fit any hypothesis. To work out the rules needs instead a series of independent tests of related species with different habits. Does the life of the praying mantis say anything about that of Don Giovanni? Are we sexual apes or angels? Without a frame of reference it is impossible to say. Fortunately, nature has given us a great experiment that might provide the answer.

Plato defined *Homo sapiens* as a two-legged animal without feathers, and – although few people see their lives as illuminated by chickens – birds are excellent raw material for the student of maleness. Centuries of work have uncovered ten thousand species, with huge variation in form and size, from hummingbird to ostrich. Their patterns of relatedness are well understood and more is known about their habits than of those of any other wild creature. Birds live in deserts, jungles and mountains. Some live as isolated pairs, others in small clusters or in colonies of thousands and they may indulge in lifelong fidelity or in wild abandon. Some males are flamboyant (and there is still room for improvement, for fanciers have selected the onagi-dori rooster to have a twelve-foot tail), but some partners are so similar as to be told apart only by a DNA test while, now and again, roles are reversed, with drab males and showy members of the opposite sex. Darwin himself turned to the avian world in his search for a general theory of reproduction and *The Descent of Man* devotes four chapters to their habits. Man is, *pace* Lady Psyche, not a goose: but geese and their relatives may have a lesson for those who hope to understand his behaviour by putting it into context.

Plenty of birds have a life-style like that of the sea elephant, without the blubber. For them, the notion of ardent males and fastidious females is simple, congenial, and more or less correct. Peacocks put their efforts into a showy tail and fly off as soon as they have fertilised their mate. In the same way, male bowerbirds build elaborate houses, decorated with trinkets. Their partners impose the bowers upon males as, trapped inside a secure display cabinet with the other party on the far side of the bars, he has no chance to force her into unwanted sex.

Complex and expensive as bird courtship might be, sperm is in most cases handed over in a few unromantic seconds. Those who donate it may then guard their partners against intruders. The skylark, for example, follows his mate at all times and warns off the opposition as he pours out his 'full heart/In profuse strains of unpremeditated art'. As no other swain can get near, he needs to mate but once or twice for each clutch. The goshawk – a bird of prey – has, in contrast, no choice but to leave his consort for hours each day to hunt for food. To ensure that a stranger has not visited while he was away, he mates again and again on his return. Each batch of eggs demands five hundred copulations.

Most birds have, like sea elephants, more variation in male success than in female. In some, such as the sage grouse, the imbalance is clear, for the participants gather on a parade ground, or lek, and the gaudy males strut about until one of them mates with most of the onlookers. In other species, loyalty rules. Shakespeare's 'coupled and inseparable' swans are famous for their devotion, and albatrosses and penguins, otherwise quite different, have the same habit.

In a final twist to the tale, a few avian societies go in for role reversal. In spotted sandpipers the female is an honorary male: larger, more aggressive and showier than the other party. As Darwin noted, in most birds males arrive in spring before their partners and squabble over territories ('Mr. Swaysland of Brighton . . . has never known the female of any species to arrive before their males. During one spring he shot thirty-nine males of Ray's wagtail (*Budytes Rai*) before he saw a single female'). In spotted sandpipers, in contrast, the distaff side turns up in the hills well before anyone else. Female testosterone levels shoot up as the days get longer, and while her mate builds the nest she – after a brief pause to lay eggs – flies off for good. She hunts for another source of sperm, and may have five clutches in a season while each male, with his paternal tasks, is limited to one. The females of some tropical species even kill the young of a broody father to persuade him to copulate with them. The coucal, a South American bird forced into this ignominious volte-face, is so unmanned by its experiences as to have but a single testicle.

Such reversals in the power struggle may seem extreme, but the fate of the coucal reveals the sad truth about how much male birds are in thrall to their mates. Their strategies are often subverted by the other party, who is much less submissive than she seems.

Paternity tests have uncovered the guilty secrets of United States Presidents and of many other men. They show also that certain birds go in for furtive copulation on a gargantuan scale. Much as the male seems to be in charge the sexual boot is in fact on the other foot, for the molecules show that a supposed father, who works hard to care for the chicks, is quite often not the true parent. Three-quarters of the hundred and more species tested show this pattern and, in some, three-quarters of all eggs are sired by an individual who does not live in the marital home.

Americans like to build nest boxes. The purple martin flies up from the south in spring, pairs up and raises its young. Some artificial colonies have dozens of rooms and hundreds of birds. At first sight, they are models of the Puritan ideal, a vast and cooperative avian township.

Genetics reveals the faults in America's backyard metaphor.

The social habits of the purple martin are quite unlike those of the pioneers of the New World. The first male to arrive advertises his tenancy and attracts a mate. A quick DNA check shows the eggs to be his own. Soon, more pairs appear, and live in harmony – but the genes cannot lie. The earliest bird is the father of more than half the later offspring as well, and most of the colony's males have no biological tie to the chicks upon whom they lavish so much attention.

When it comes to infidelity, the other party is often in charge. The superb fairy-wren, an Australian bird whose plumage lives up to its name, is the most promiscuous yet found. Courting males pluck a bright flower and show it off against their own cobalt-blue plumage. The objects of their attention seem impressed, but they are happy to cheat. Many make a beeline to a paramour's nest, just before sunrise, ready for a quick affair. On the way they pass through the territories of several less desirable birds, as proof of who makes the decision to indulge in furtive copulation.

Superb fairy-wrens may be devious, but other birds appear to be more charitable in their ways, as several individuals join forces to raise the young of another female. Why do they do it? Helping at the nest, with its social and political overtones, has been studied in dozens of species. The reasons (or, at least, the theories) behind such apparently altruistic acts are endless, and some, no doubt, are accurate. In their diversity they hint at how what seems a simple end can be reached by complicated evolutionary means.

In bees and wasps some animals give up on their own genetic future. They pass on their DNA in an indirect way by aiding their relatives, who carry a version of the same thing. Sometimes (but by no means always) such logic works for birds. Young female Seychelles warblers, for example, help their parents and sisters feed the young, which improves their own genetic prospects. Quite often, though, helper and helped show no particular kinship. The Galapagos hawk –

noted by Darwin on his visit to the islands for its bold and tame behaviour – is happy to act as nursemaid to anybody, while bachelor scrub-wrens prefer, in a rather perverse fashion, to assist unrelated birds. Certain birds go in for political, rather than biological, alliances. The hedge sparrow was referred to by a Victorian vicar as a paradigm of what his flock should aspire to, but their own dear Queen would not have approved of its habits. They live as groups of unrelated individuals. No foreign male is allowed in, but within the tribe promiscuity rules, with half of all young fathered by someone other than the main occupant of the nest, and plenty of free support from birds who have no genetic investment in the young.

Those who lend a hand might gain as they learn to be good parents, or the habit may allow them to join a colony (useful when enemies are about) or give them the chance of entry into a landlord's domicile when he dies, or of an erotic encounter with his children. In some groups, history is more important than are the sexual pressures of today, for lots of crows and their relatives are great helpers, although they are otherwise quite distinct in their ways. Presumably, some ancestor evolved the habit and it stuck.

Among the birds parental care, cheating and helping by both mothers and fathers have evolved for many reasons, most of which we do not understand. The avian world in its variety is a miniature of maleness. It has lessons to suit all tastes: from peace to passion, from fidelity to vast profligacy, from honest signals of a potential father's worth to a hidden system of choice by his mate (with a bit of transvestism thrown in). This genital gallimaufry evolved in a group which appeared before the mammals and has habits even more diverse than theirs. A century of ornithology and a decade of genetics have uncovered more about the secrets of dozens of kinds of wild birds than of a single wild mammal. Surely, if any group has the key to masculinity it must be those two-legged upright vertebrates with feathers but, alas, they do not.

Birds tell a tale instead of infinite opportunism slightly constrained by history. In spite of some vague general hints that fast-living species, with high rates of birth and death, are more promiscuous, their behaviour shifts in an expedient and uncertain way. Sometimes it alters so quickly as to overwhelm any attempt to find the rules but sometimes an ancient habit stays on in animals that have otherwise undergone many changes. Close relatives are often quite unalike. Two populations of the same species of bowerbird a hundred miles apart in Irian Jaya differ in male behaviour; in one, bright ornaments are chosen while the others prefer dull pieces of bark. All ducks have penises, rare among birds, and some go in for forced copulations that may drown their mates (a habit referred to by the credulous as 'rape'). The male organ of the Argentinian lake duck, a bird which weighs a mere pound or so, is a corkscrew-like structure a foot long when erect. The creature is wildly promiscuous, and its remarkable penis evolved to displace the sperm of earlier mates. In contrast, several of its relatives, armed though they are with more modest versions of the same thing, stay as stable pairs. Some garden birds are faithful and some are not and some — such as the hedge sparrow (which lived on Alpine pastures before it moved into gardens) — have inherited the habits of an ancestor and have not evolved such behaviour for reasons of their own.

The avian world is a warning to those who search for a theory of manhood. To compare close relatives is a staple of all evolutionists, but for sex, the most polymorphously perverse of all pastimes, the pace of change and the power of history mean that the method often fails.

Men, famously, share around 99 per cent of their DNA with their two closest relatives. Perhaps our primate kin hold the Freudian key; but, perhaps, they do not.

The primates appeared about sixty-five million years ago, before the death of the dinosaurs. They have split several times since then; into lemurs, monkeys and apes, and the apes themselves into dozens of extant species, men, chimps and gorillas included. Chimpanzees broke away from our own line around five million years ago, with the last ancestor common to humans and gorillas perhaps two million years earlier, and to the orang-utans a few million years before that. What do our relatives say about the evolution of male behaviour – our own included?

Because most are rare and hard to study, far less is known of their ways than of birds. Anthropomorphism is also a problem. Few people bond with a fruit fly (or even a goose) but those who study apes often see themselves as members of a shared tribe. Chimpanzee tea parties are notorious (F. D. Roosevelt suffered during his presidential campaign when more spectators turned up at Detroit Zoo than at his political rally) and even scientists are not immune: they name their animals – Humphrey, Greybeard and Willy Wally, their mates as Flo, Gigi and Godi – and are delighted by Gigi's generosity and shocked when what seemed a gentle father kills her young.

In spite of this desire to humanise the less than human, we have learned a lot about the reproductive lives of our relatives. As in birds, sex brings forth diversity. Primates go in for most conceivable patterns – monogamy, one female with several males, one male and numerous partners, and so on. The lek alone is lacking.

Most primates live in groups (even if none apart from ourselves exists in numbers close to those of a seagull colony) and baboons may form tribes of hundreds of individuals. In contrast, most of the primitive and nocturnal prosimians live solitary lives, except when they mate.

Primates have a commendable range of erotic interests to match their varied social lives but the two do not always overlap. Solitary creatures such as the indri of Madagascar get a chance to mate only now and again, while certain macaques who live in large packs may mate forty times a day. Plenty of primates are monogamous. The hamadryas baboon hangs round tourist sites in North Africa and can be a pest. To shoot baboons attracts complaints, and the park managers control their numbers with castration instead. It works well. Monogamy survives and the females remain unfertilised. The habit has evolved several times in different conditions. The lonely indri goes in for it, as does the lar gibbon, which lives in groups.

More kinds are promiscuous. As female mammals (unlike birds) do not in general store sperm, the last male to mate gets a great premium. They strive for that status with a variety of tactics. Some group-livers have a dominant male with several consorts, but in other social creatures, such as the red-tail monkey, animals who wander in from outside do their share. In some species, a female is accompanied by two mates, in others a single patriarch controls several partners, while in a certain marmoset, several faithful pairs live as a harmonious assembly. In saddle-backed tamarins, in contrast, several males share a single mate, and in some langurs females seem to be in charge as they pursue their opposite numbers in search of sex and copulate repeatedly with different animals. This might confuse their partners as to who actually is the father and reduce the chance of their killing another's children.

Primate bodies vary almost as much as their habits. Some males are hard to distinguish from their partners, while others are huge in comparison, or fight with giant canines. A few signal their power in less violent ways. Lemurs assert their presence with a territory marked out with pungent scent while howler monkeys do the same with sound. A mandril uses his blue and red face, a proboscis monkey depends on his bulbous nose, while the guenon turns to his famous red, white and blue display (red penis, blue scrotum, white pubic hair).

Perhaps the male member itself is driven by competition, as longer versions deliver closer to the target, or perhaps its recipient's choice of some arbitrary erectile attribute is important. Whatever the cause, monogamous penises tend to be small (the faithful owl monkey has a phallus no more than a fortieth of its body length when limp) and promiscuous species tend to have the largest organs. Almost all primates have a baculum, a penile bone – and it evolves with great speed. Those of solitary species like the galago are longer than most, while the penile bones of Old World monkeys dwarf those of their cousins across the Atlantic. When it comes to the testes, some of the wanton kinds go for quantity in the spermatic race, but the faithful owl monkey makes minute amounts of sperm in tiny testicles.

This diversity of tactics allows individual primates to be used as evidence for almost any conceivable pattern of sexual life. The lesson from our relatives is not much clearer than that from birds. Big males with bright colours, sharp canines and generous genitals do hint at sexual battles but the story is not always simple. The canines of some male baboons are four times longer than those of their mates, but the teeth of tarsiers, creatures just as keen on fights, are almost the same in each. Male tree-dwellers, however avid in the hunt for mates, are relatively smaller than those who stay on the ground, as a hint that other forces (gravity included) limit their ambitions. Mammals as a whole show a strong association between the differences in the size of the sexes and the extent of competition for partners, but primates do not.

Perhaps the problem lies, as in birds, in the very diversity of the group to which we belong. Monkeys and apes live in deserts and jungles, in Old and New World, and vary in body weight by hundreds of times. Given the speed at which sex evolves, to search for the remnants of any fundamental rule in such a miscellaneous group may be to ask too much.

We have some closer relatives. Perhaps the key to man's

behaviour, at least in its untamed state, lies in the almost human trio of chimpanzee, gorilla and orang-utan. Their brains are smaller than ours, but they contain a similar set of genes (although more are turned on within our own skulls). In spite of such affinity, their erotic habits are remarkably diverse.

Chimps – the genus *Pan* – get their scientific name from their interest in copulation. Their society turns on a set of uneasy political alliances within a larger and more stable group. Males form loose alliances of a hundred or more, which, like Montagus and Capulets, represent loose bands of kin in endless warfare with the neighbours. Internecine battles are quite mild, but those between groups can be lethal. Males settle down as groups of friends that – rather like the Athenaeum – are based more on age and status than on kinship and whose members, like those of a London club, build their lives on mutual grooming and shared meals.

Paternity tests show that few fathers come from outside the tribe. Within it, the top males tend to win, with the dominant animal the parent of half the young. Life can be hard for that fortunate chimp, for he has more testosterone and stress hormones than do his underlings. His rule is always under test, and he is kept busy with fights with his rivals. Kingship does not last long.

Sometimes several chimpanzees mate with the same female, but now and again one manages to keep his consort to himself for weeks. A captive animal may copulate every five minutes, which depletes even his generous reserves of sperm. In the wild each female mates a thousand times for each pregnancy. Chimp penises and testes are impressive, as a further hint of spermatic competition (and if men were built on the same scale, their testes would be the size of grapefruits). The famous animals in the Gombe reserve are lusty and aggressive. They kill outsiders and are fierce in defence of their mates, but much of their behaviour depends on the bananas provided by the scientists who study them. Unfed groups near by are less clannish and more ready to accept intruders.

Chimpanzees have a close relative who hints at the flexibility of ape habits. The bonobo, or pygmy chimp (which is so similar to its cousin that some biologists do not see it as a separate species), is a much more relaxed creature. Its DNA dates its origin at less than a million years ago, long after the departure of the human line. Life in its homeland around the Zaire River is easy, with more food and less competition from other primates than in places such as Gombe. Females, rather than their consorts, form social groups. They live in large troops and are much more open and peaceful in their behaviour. The bonobo has a libidinous life, with lots of face-to-face copulation, group sex and plenty of homosexuality. The animals copulate more or less all the time and not just when the females are on heat.

The behaviour of the gorilla stands in stark contrast to the bonobo's happy world. The animals are social, but their lives are unrelaxed. They roam around in gangs of twenty or so, but the top males spend more of their time in fights than in copulation. One individual is dominant, and keeps his status for several years. As females are receptive only now and again, he can afford to take it easy for much of the time, and leaps into aggression only when a junior attempts to mate. DNA tests show that – just as in the cheating fish of American lakes – his supposed subordinates do just as well in the contest so that his impressive efforts are wasted.

The orang-utan has habits of its own. It spends friendless months in the jungles of Borneo, much of the time up in the trees. Solitary as the animals are, sexual competition is fierce. Males are twice as big as their opposite numbers and winners hold large territories whose boundaries shift from day to day. Their mates have their own patches, but these are small, and a patriarch has several partners within his kingdom. He spends hours calling to advertise his presence. The most charismatic males of all grow a broad face, which scares off the opposition and recommends itself to potential mates. When copulation is over the long period of solitude begins again. Young orang-utans suffer great social stress in the hunt for a mate. In captivity, an adolescent animal develops his long hair, cheek flanges and typical bellow by the age of ten. Put him close to a more dominant individual and puberty is delayed for seven years. In the wild, such striplings may sneak up on a female and force copulation upon her. She flees for protection to the nearest major male.

Our relatives have different life-styles, and our ancestors, too, were distinct. The ancient fossil predecessors of humans, chimps and gorillas differed more in size by sex than does any primate alive today. On the line to ourselves, canines began to shrink, but males stayed much larger than their opposite numbers. Even our immediate ancestor, *Homo erectus*, may have had males much bigger than their partners (the story is confused by the possible existence of two species, one large and one small).

Nowadays, most people do not behave much like orang-utans, but chimpanzees and perhaps even gorillas strike a certain chord. But what can we really learn from a comparison with our close relatives, alive or extinct?

Men and women differ less in body size than do the sexes in any of our kin. Our canines are much the same in the two (the claim that teeth were supplanted by weapons does not hold water as fangs got smaller before the invention of tools). We do rather better in the genital department. When it comes to penis size, most apes are not well endowed and even the mighty gorilla has a half-inch penis bone and testes of just five-thousandths of its body weight. Those of the chimp are, in relative terms, ten times as big, and its delivery organ is impressive. Our own weigh in at a mere twice the gorilla level, but are accompanied by a penis which stands far above either. Copulation, too, is a less perfunctory business in *Homo sapiens*. It takes seven seconds for the chimp, fifteen for bonobos, and a minute for gorillas. We are the most relaxed, at an average of four minutes from penetration.

Quite what to make of all this is hard to know. Man's increased height, reduced life expectancy, longer nose and shorter temper in comparison to his partner have all been blamed on his sexual state. Even baldness (shared by several male primates) has been ascribed to sexual selection. Perhaps a hairless head increases the apparent size of the face so that when a bald (and hence older and more important) man goes red with anger he scares off the opposition (which is odd given the rival claim that a bristling beard – which hides the face – has the same effect).

Speculation about sex is too easy, but our frame does retain some hints of ancient times. *Homo sapiens*' generous testes suggest a history of moderate promiscuity and his large phallus might also be a clue to a salacious past, but man's modest teeth and the similarity in body size of each partner point at a certain fidelity. The figures are confused because human height and weight (but not, presumably, the dimensions of our most intimate organs) have increased in the past few centuries. Some less direct clues come from local patterns of change in the Y chromosome and in mitochondrial DNA (inherited through females). The geography of the two lineages is only moderately different in humans, but enormously so in chimps – a hint, perhaps, that promiscuity, with a dominant male ruling a large territory, has been less of a force in human history than in that of our closest relative.

In terms of evolutionary affinity, the lives of the four great apes make little sense. The bonobo and the chimpanzee split long after chimps and humans but they differ markedly in behaviour. Gorillas are further away from us but are faithful and copulate less than either kind of chimp, while orangutans are quite unlike any of their relatives. Our own societies, too, are quite diverse in their ways, although the general picture in the world today is of reasonable fidelity, with occasional lapses by both partners.

Biologists, from Darwin on, have often tried to put their subjects' sex lives into context. In many creatures, the balance of cost and benefit as related species with different habits struggle to pass on genes gives a new insight into how maleness evolves, but so far the logic has not worked well for the great apes. They do little more than confirm what we already know: we are fairly faithful primates, whose behaviour changes with circumstances. Perhaps *Homo sapiens* once danced to the same erotic tune as did his relatives, but it is hard to work out what the melody may have been and whether it has much relevance to the genetical gavottes of today.

As the shortage of sperm donors shows, men are different in another and essential way. For them, sex relies on more than a mere series of sums based on investment and return. A recent account of the evolutionary economics of childbirth has made the disquieting claim that 'In short, there is no explanation for why Americans still want children.'

What is true for Americans applies to much of the world – but copulation flourishes. Man's continued interest in the pastime does suggest that biology plays a part in our lives, but his disregard of the rules of the erotic market place illustrates the agreeable truth that men, whatever their secret desire to act like chimps, are not altogether bound by biological laws.

Karl Marx put it well: 'Eating, drinking and procreating are . . . genuinely human functions. But abstractly taken separate from the sphere of all other human activities and turned into sole and ultimate ends, they are animal functions.' Men are forced to eat and drink, but, unlike birds and apes, can, quite voluntarily, separate procreation from their other activities. As a result, the main lesson from nature is that, for ourselves alone, sex does not make the world go round. Or, in the more economical words of Mark Twain, 'Man is the only animal that blushes – or needs to.'

CHAPTER 10

A MARTIAN ON VENUS

The Urinette Company in the end went bust. They had hoped to improve the efficiency of public toilets with a receptacle into which women could urinate while upright. A rigorous research programme showed that this was both physically possible and more hygienic than the traditional lady-like posture and the She-Inal went on the market. Impressive as their business plan was, the product did not catch on and was soon withdrawn.

Urinette's mistake was to fly in the face of common sense (and in Texas, where a law against female urinals was passed in 1997, to outrage all norms of society). Everybody knows that men make water while standing while women prefer to squat. Men also tend to be fonder of violence, pornography and barbells than their partners. But why? How much of the difference, from micturition to masturbation and murder, is inborn and how much is imposed by society? Some male talents, from lifting weights to vertical urination, may come from the constraints of his frame, but some – even homicide – could be a response to an evolved sexual imperative. To understand the truth might explain, if not excuse, man's behaviour; but the truth, even for urinals, is hard to pin down.

Genes are of course involved in making water, as the penis, the fruit of the Y chromosome, is if nothing else a convenient conduit for waste. Such behaviour, however, is not firmly coded into DNA as boy and girl babies both void with the same careless abandon. Culture is much involved. Young boys born with ambiguous genitals often ask for surgery so that they can stand up at a urinal like their friends. Orthodox Jews, in contrast, teach their sons to micturate with hands free: 'Better', they say, 'a bad aim than a bad habit!' In Arab societies men crouch while, if the strangely detailed account of the sexologist Havelock Ellis is to be believed, French women were once happy to perform the ritual while upright. A check on different creatures (a staple of biologists when they search for genes) is not much help. Like ourselves, dogs do it from the vertical while bitches lower both hind legs, but in cats males and females behave in much the same way.

For urination, common sense is not much help in distinguishing nature from nurture – but neither, so far, has been science. For many other talents associated with inborn differences in the lower anatomy, most people are in accord: the sexes are born unalike. John Gray, in his *Men Are from Mars, Women Are from Venus*, goes so far as to claim that the two halves of society are 'from different planets, speaking different languages'.

Mars – an airless world with an icy past which dried out to leave an arid and infertile landscape – makes a handy metaphor for male behaviour. Whole areas of man's life appear to be fixed by gender and many are repelled by any deviation from the norm. In the 1955 film *Rebel Without a Cause* James Dean seeks out his father for advice on his adolescent problems, but recoils in disgust when he finds him in an apron, hunched over the sink. Society is more relaxed than it was in that dismal decade but still has an almost automatic tendency to believe that men and women are quite distinct in their approach to life. A special chromosome is, as a result, a useful alibi for man's excesses.

The structure is often called upon, as a quick count of the numbers of each sex in prison shows. For the extremes, from patterns of crime to those of mental illness, nobody denies the existence of inborn differences between people with and without a Y. Even so, it is not easy to say how far DNA makes the average man behave the way he does. Whole libraries of books set out to find his inner self and his supposed tendencies towards violence, greed and self-doubt. Unfortunately, much of the literature on the subject is on the margins of science (and, for that matter, of literature). A third of his brain, says Robert Bly's bestseller *Iron John,* is a warrior brain; and half the population needs to return to a time when 'the divine . . . was associated with mad dancers, fierce fanged men, and being entirely underwater, covered in hair'. Men can, Iron John suggests, find their primal selves by group drumming.

The notion of modern man as the savage tamed is everywhere in *The Descent of Man* and became universally popular soon after it was published. In the United States, a century and a half ago, the Improved Order of Red Men had thousands of members who took part in Native American chants. The nation still has binges of bonding. Bohemian Grove, in the California redwoods, invites male millionaires and politicians to an annual celebration of masculinity. Even President Nixon, himself once a Red Man, came. He stayed in a lodge called Cave Man Camp.

Such claims seem odd to those of us who do not come from some metaphorical Mars. We cringe at the idea of percussion, alone or at camp, and are content to leave our warrior brain, if it exists, in peace. The Y is much appealed to as an excuse for masculine excess, but most of the evidence comes not from science but from vague social speculation. John Stuart Mill made the case in his *Principles of Political Economy* of 1848: 'Of all vulgar modes of escaping from the consideration of the effect of social and moral influences on the human mind, the most vulgar is that of attributing the diversities of conduct and character to inherent natural differences.' He contested the claim that the Irish were of their nature indolent and pointed out that in England they worked as hard as anybody else. Only at home, where the system of land tenure meant more rent, rather than more food, for hard work, did they take it easy.

Mill's argument applies with equal force to sex. In his day the notion of a bodily structure as the key to identity would have seemed strange. Nationality and class each took precedence over genitalia. Females might be of lower status at all levels of society, but a plebeian man saw himself as inferior to a woman of higher station (and she tended to agree). Not until a century later, with the growth of science, did the notion arise that the body alone is real and that those who inhabit it are constrained by biology rather than the social order.

Attitudes are inherited as readily as are genes (even if they evolve a lot faster). Children's books show how quickly things can change. They once dealt with the young not as boys and girls, but as infants well brought up or otherwise. In *Tales Uniting Instruction with Amusement* of 1810, Tom Tindall, who has killed his mother with a firework, 'wishes he had followed his poor father's good advice. If he had done so, he might now have been at a genteel boarding school, with both his eyes safe, instead of being a chimney-sweeper, and blind of one eye'. His sister is just as naughty. She swallows thread, against instructions, and dies of strangled intestines: 'From that moment the children unanimously agreed strictly to attend to their father's orders, and never in the slightest instance to act in opposition to his will.'

Fears of social upheaval gave place to a terror of sex. For the Victorians, men and women became different species; angels and – at best – frightful insects. Heroines were condemned to a romantic and convenient death, while their brothers found a destiny in foreign parts. As the century progressed, so did masculinity. In some libraries, books written by men were put on separate shelves from those by women.

Games were *de rigueur* ('All boys who like to participate may participate, and all others *must*') and fiction followed. G. A. Henty (author of *At the Point of a Bayonet, Won by the Sword, Sheer Pluck* and other instructive works) complained to Robert Louis Stevenson that the latter's famous books for boys needed less psychology and more claymores.

The position of each sex was fixed. Even their cells behaved in different ways. The egg, said Stevenson's contemporary, the biologist and reformer Patrick Geddes, was passive, conservative, apathetic and stable, while sperm were active, energetic, impatient and impassionated. If sperm and egg had separate personalities, what could be expected of those who made them?

Claymores can certainly take control when they get the chance. The *Bounty* mutineers reached the island of Pitcairn in 1790. Their ship bore nine British sailors, six men of Polynesian origin, and thirteen Polynesian women. By the time the colony was discovered in 1808, just a solitary founding male was left. Twelve of the original men had been murdered, one had committed suicide and another had died a natural death. Ten of their thirteen partners survived.

Such tales, at first glance, support Henty's view of manhood as destiny. His notion is still popular (even if its advocates lack his literary flair) but a closer look casts doubt on the whole notion of biology as fate.

Warrior brain or not, men and women are, in general, much the same. For food, sleep, conversation and more, people have their own preferences, but most of them reflect variation among individuals, without much respect to sex. Fewer husbands than wives are anorexic and more are obese, but the presence or absence of a Y chromosome is not central to the average citizen's choice of restaurant, or his or her waistline.

Sex suffers from simplicity. With just two categories it becomes fatally easy to see opposites. As a result, many traits, from violence to depression, are often discussed in its context. Experience may be just as important but the stark distinction offered by manhood is hard to ignore. A 1936 survey had maleness at one pole and its opposite at the other. A positive response to 'Do you rather dislike to take your bath?' was an affirmation of masculinity, but to know the answer to 'Things cooked in grease; are they boiled or fried?' was a grave challenge to a boy's sense of self.

Patrick Geddes, he of the apathetic egg, agreed. To him 'a deep difference in constitution expresses itself in the distinctions between male and female . . . What was decided among the prehistoric Protozoa cannot be annulled by Act of Parliament.' Some still see the influence of history in men's lives, while others, repelled by such fatalism, wish away the phallus and argue against any hint of divergence between the sexes, whatever the evidence. So fixed is each belief that it can be hard to find the truth.

Many of the supposed contrasts between mental planets do fade on further scrutiny. Men, as countless works of self-improvement assure us, are more concerned with status, more independent and more dominant than are their partners. Obvious as those claims might seem, objective tests show gender to play but a small part – a tenth or less – in where anybody sits in the spectrum. Class is important, as is culture, with whites, for example, seen as more passive than blacks. Among the patriarchal Mormons, marital arguments often end with the husband as winner. Among the more egalitarian Texas farmers the result tends towards a draw, and in Navajos the wife (around whom the household revolves) often comes out on top.

Martians and Venusians do not speak different languages but have suffered a mere shift in dialect. Women use more redundant phrases ('Men are more powerful, *aren't they*?') but these can show confidence rather than submission ('*I think* it's time for a change'). The British, at least, find it easier to establish class, rather than chromosome, from the use of language.

Expectation is important in men's lives. Pairs of students, assessed as bossy or meek, were asked to remove a bolt inserted into the wall of a large box, with the nut on the outer surface. One had to go in and hold the head steady, while the other had the more difficult job of removing the nut with a spanner. When two men worked together, or two women, the aggression test predicted who would lead. A meek man paired with a bossy woman got to unscrew the nut – until the couple had talked for a few moments and had established their social position. Then the male, aware of his place, crept inside.

Society may force men into boxes, but the role of sex in their affairs cannot be denied. It becomes most obvious when the subject itself is on the agenda. On American campuses men find 'communication' the main problem on a first date, but those at the receiving end complain more about 'unwanted pressure to engage in sexual behaviour'. When the students were asked how long they would need to know somebody before asking to sleep with them, the males averaged a week (and plenty were happy with an hour), while their prospective mates felt that six months was more appropriate.

The Descent of Man has a long account of how such differences in erotic attitude – crass male against fastidious female – has moulded evolution. Preference for a particular type of mate not only drove the emergence of *Homo sapiens*, but led to the contrasts in appearance of, among others, Africans and Europeans. As Darwin pointed out, most differences among populations are small (he was 'impressed with the close similarity between the men of all races in tastes . . . by the pleasure which they all take in dancing, rude music, acting, painting, tattooing . . .') but they look distinct. The world map of human DNA does indeed show that the average divergence among individuals – two Ghanaians, or two Icelanders – is several times greater than that among the so-called races of humankind. Even so, they do deviate quite strikingly in appearance. Sexual selection drives such changes in birds, so why not in ourselves? Darwin certainly thought so ('If it can be shewn that the men of different races prefer women having certain characteristics, or conversely that the women prefer certain men, we have then to enquire whether such choice . . . would produce any sensible effect on the race') and *The Descent of Man* makes great play with the fact that groups do vary in their erotic preferences. South Americans insert metal studs into their lips, a habit then unknown in Europe, and the Chinese think the Europeans hideous (a view once reciprocated by Europeans). Such shifts in sexual tastes might, given time, cause whole populations to alter.

Most modern biologists see the evolution from black to white or from round to slanted eyes as a result more of climate than of erotic preference, but Darwin's idea is feasible, albeit hard to test.

As sociobiologists and authors of romantic fiction have noticed, humans exhibit mate choice. Quite who chooses whom, and how, is not altogether certain. Rigorous research by novelists and others has discovered that men prefer young and beautiful women and that the attractions of the elderly increase if they happen to be millionaires. Beyond that (and the genetic basis of plutocracy has not yet been established) the role of sexual choice in the evolution of human diversity is not at all clear.

In the developed world, men and women – whatever their colour – tend, when given photographs of potential partners, to prefer those relatively light in hue. That is true both for blacks and whites, which militates against Darwin's idea of selection by erotic choice; but a modern society that emphasises the beauties of blondes may have altered man's tastes.

As the genetic differences between Indian castes show, social barriers can indeed influence the decision about who to mate with, but, unless there is a tie between genes and preference, this will not cause populations to change. Sheep have a drive to copulate with someone who looks like their mother, and to a lesser extent the same is true for men. Perhaps in ancient times local partiality did favour a particular physical type. Race alters our perception of difference, as people find it harder to recognise photographs of others from a group other than their own, but the evidence that this drives sexual tastes is hard to find. Such barriers are now fragile indeed. In colonial days, European men did not care where they spread their genes and in these kinder times the impediments to interracial sex are also feeble, with about a third of black British women having a white-skinned partner. If sexual selection once drove evolutionary change, the age of cheap air tickets has much reduced its power.

Some useful talents do favour males. Imagine an empty pint glass tipped at forty-five degrees. Now think of it as half filled with beer and draw the line at which the surface of the liquid settles. Most men draw it, as they should, as parallel to the table top, while many of their partners expect the beer to stay in line with the top of the glass. Now ask people to match musical notes or to rhyme words. This time proportions are reversed, with more correct answers from women. Men remember simple pieces of information such as phone numbers less well than do their spouses, but when it comes to manipulating complex three-dimensional images in their heads, they do better.

As they learn to read, boys can feel like fish out of water. They are baffled by the lack of fit between how a word is written and how it sounds. 'Fish', after all, can be spelled 'ghoti' ('gh' as in rough, 'o' as in women, 'ti' as in nation) and so, more or less, can 'water' (this time with the same letters but as in plough, cord and tin). A child with fluent speech may be plunged into an alien element when he starts to read. Dyslexia – an inability to match written and spoken words – is twice as common among boys than among girls. English has forty sounds spelled in a thousand ways, while Italian has just twenty-five, with a mere thirty-three ways in which to put them on paper. As a result, Italy has, per head, half as many overt dyslexics as Britain. More turn up when special tests of comprehension are used, but, once again, most of them are boys.

In some creatures the sexes also get lost in their own way. Teach rats a path through a maze, and females succeed if given lots of landmarks. Remove the clues and they do worse, while their mates still remember the route they took on an earlier try. Voles navigate for more romantic reasons. Prairie voles are faithful creatures who stay at home for most of their lives. Their meadow vole cousins are more fickle and the males wander the countryside in search of new partners. Male meadow voles find their way through a maze much better than do their mates, while in the more matrimonial prairie type both parties are just as bad at the job. In humans, too, hormones might be involved, for women who undergo testosterone treatment before a sex-change operation get lost rather less easily than before.

More men than women become scientists, and in the United States boys outnumber girls by seven times in the top 1 per cent of children who shine in the subject. The advantage shifts the other way when it comes to literary skills. Among hundreds of occupations in scores of pre-literate societies, metalworking is the sole craft confined to men. It demands the spatial skill and the strength that are each, some say, manly attributes. In chimpanzees, in contrast, tools are used more by females, so that mechanics is not always coded for on the Y chromosome. Quite how fixed such traits might be is also not clear, for the gap in mathematical ability has halved over the past twenty years. However, the difference in the capacity to handle three-dimensional objects has not.

The mere existence of a difference between men and women does not, needless to say, prove that it is inborn. To do that needs subjects who have not yet learned what is expected of them. Childhood – at least in its fictional form – has moved on from Tom Tindall. Modern parents in their urge to be politically correct tend, unlike their fore-bears, to see their infants as experiments in nurture in which nature plays no part. From *The Wind in the Willows* to *Winnie the Pooh* infancy was almost asexual and a neutered kind of boyhood became the norm. Even girls were androgynes. George (a young lady) in Enid Blyton's *Five on a Hike Together* rails against her fate: "'It's stupid being a girl," said George, for about the millionth time in her life. "Always having to be careful when boys can do what they like."'

Today's mothers and fathers prefer gentle boys and con-fident girls as heroes but those who read the books go for the stereotypes (as the success of the *Harry Potter* series shows: Hermione faints at the first sign of a troll while Harry and his chums hit it with a steel pipe).

Even before they utter their first sentence, boys and girls feel themselves to be distinct. At six months, babies given a series of pictures of women to look at soon lose interest, but their attention is at once drawn to an image of a man. By two, most girls prefer dolls, and most boys, cars. A year on, almost all have a firm identity as members of their own group (even if a few are confused by the notion that a boy who puts on a dress turns into a girl). At five or so, boys accept the sad truth that they are stuck with their state and become as sexist as the most unreformed adult. Their toys deal with objects, while those of their sisters concentrate on the emotional universe (and, as letters to Santa show, not just because parents choose them). Boys play with boys ten times more than with their opposites; they prefer larger groups, rougher pastimes and more time outside. Their games take longer because they argue about the rules and, if forced to talk to a girl, the subject soon turns to the urgent need to find another way to spend the time.

Some of this may well be in the genes, but parents – even if they try to be unbiased – also mould their children's future in a way that has little to do with DNA. Faced with a baby dressed in an ambiguous manner, adult attitudes depend on whether they believe it to be male or female. A supposed boy is less often given a doll to play with, an alleged girl is kept safe from toy tractors. Make the unlucky child cry and those who believe it to be a boy diagnose 'anger' rather than 'fear', while those told otherwise draw the opposite conclusion. Parents handle sons more roughly than they do daughters, and are convinced (against the evidence) that the former have a stronger grip.

A young brain responds to such outside pressures, but sexual identity is also imposed by the internal world. Important as experience can be, the contents of a boy's skull are influenced by messages, both direct and indirect, from his insidious Y chromosome. From his earliest days his brain, like his body, is to some degree under its control.

Often the best evidence of how a piece of biological machinery works emerges when it goes wrong. To an extent the approach is helpful, but to use disease as an insight into normal behaviour involves a certain risk. Cars – like bodies or minds – often break down, but their day-to-day perform-ance is not much related to what happens when things go awry. Even so, the errors of sex give some insight into how the two models of *H. sapiens* function when they are running well.

Saint Wilgefortis, daughter of the King of Portugal, grew a beard in protest when betrothed against her wishes to the King of Sicily. The Sicilian at once refused her, and her father had Wilgefortis crucified. The Reverend Alban Butler, in his *Lives of the Saints*, describes the tale as having 'the unenviable distinc-tion of being one of the most obviously false and preposterous of the pseudo-pious romances by which simple Christians have been deceived or regaled'.

Her story is, nevertheless, probably true, at least in its clinical details. The adrenal gland, above the kidney, has several tasks.

It helps control the body's state of alertness, regulates its salt balance and makes steroid hormones. Its products are involved in the response to stress and in sugar and salt regulation, and have the power to induce maleness. If the gland is damaged by a tumour, an excess of testosterone-like substances is made. Women can then suffer unfortunate symptoms, beards included.

Inborn errors can lead to symptoms like those of Wilgefortis. If not enough secretions are made, the body pushes the adrenal gland to work harder – and, as a result, to make more masculine hormones. In boys this may switch on puberty too soon, and in girls can lead to a variety of troubles, from acne to abnormal genitalia. Some children with the condition are unable to control the amount of salt in the body, and die unless treated.

The inherited version of the virtuous but unfortunate lady's illness is known as congenital adrenal hyperplasia and is by no means rare. Together, the inborn errors of the adrenal gland represent the commonest set of single-gene disorders (apart from those such as colour-blindness, borne on the X chromosome, which show themselves in all boys who carry them). Among Hispanics in New York one child in fifty is born with some damage to the gene involved. Such problems can cause a fetus – even in the absence of a Y chromosome – to develop, more or less, as a male. The genitals are masculinised, and the child may have more hair and a deeper voice than normal. An XX fetus with severe injury to its DNA may develop as a boy, as a girl, or with the physical attributes of both. In some cases the condition can, if diagnosed early enough, be treated before birth with a drug able to cut down the gland's activity, but often the illness is missed until too late.

Severe cases of congenital adrenal hyperplasia are rare – but more than fifty different mutations with a milder effect are known. Some – such as the version found among the Sambia tribe in Papua New Guinea – do not make their presence felt until puberty, when a supposed girl begins to manifest the presence of a penis. The unfortunate sufferers ('turnim-men', as they are called in pidgin) are much discriminated against.

The steroid bath before birth can alter the body in an obvious way. It may also have a more subtle effect upon the brain. Boys show little change, but – some say – the behaviour of girls might suffer a slight shift.

A few girls with congenital adrenal hyperplasia prefer boys to girls as friends. They play with more vigour than most of their sisters and choose guns rather than dolls as toys. With maps they do better, but with babies worse. As they grow, some become interested in football, cars and so on. Almost by default, such people may turn to careers often seen as manly. Sometimes they are attracted to their own sex and a small minority decides to go through surgery when adult to give them a penis.

All this hints at the power of a brush with chemistry, but the story is – needless to say – not simple. A woman who takes up a career in a field more associated with men does not have an engineering hormone. For most of those with such mutations the effects are weak, and in a large British survey almost no behavioural effects were found. For girls with a major mutation, ambiguous genitals – which become more obvious with puberty – may play a part in their choice of partner. What is more, such children often spend much time in hospital, an experience that can push a girl's behaviour in a boyish direction. In practice, most of those who have a mutated version of the gene live their lives with no trace of difference from others.

Other creatures give firmer proof of the effects of hormones on the brain. Male birds put a lot of effort into passing on genes. Song, territories and bright plumage are expensive and some species shut down their efforts except in springtime. In winter, the testes reduce in size by a hundred times. For those species in which just the males sing a certain segment of the brain is five times bigger than in their partners and it shrinks in winter when courtship stops. The link between cause and effect within the avian skull is clear.

The notion of a gender difference in our own cranial contents once seemed obvious. One French anatomist was confident: 'In the most intelligent races, as among the Parisians, there is a large number of women whose brains are closer in size to those of gorillas than to the most developed male brains . . . Those who have studied the intelligence of women recognise today that they represent the most inferior forms of human evolution and that they are closer to children and savages than to an adult, civilised man.' When it came to Negroes, matters were even simpler as their brains were the equivalents of those of a white child, a white woman, or a white man with senile decay.

Claims of inferiority for women and Negroes are still around today, but the truth is harder to find. Innumerable sections of the male brain have been said to be larger or smaller than those of his partner, but few of the supposed differences stand up to scrutiny. Men may have a broader set of nerve connections between the left and right sides – but only in dead specimens, and not in brains scanned while the owner is still alive.

Husbands and wives do show slight divergence in certain small groups of cells within the hypothalamus, the structure at the centre of control of reproductive hormones. What is more, the masculine brain has, in some studies, rather fewer nerve cells in parts of its surface than does the feminine. The sexes may also differ in what the brain makes. Low levels of serotonin are associated with depression (an ailment more frequent in women) and men generate the substance at almost twice the rate of their partners. Whether such changes dispose to differences in behaviour, or the other way round, is not clear.

When words are flashed up on a screen visible to only one eye, women do better when the image is presented to the left side of the space (which passes its information to the right side of the brain). What is more, after a stroke on the left side, five times as many women as men lose the ability to speak. Science can now see the nervous system at work at such tasks by looking at the flow of blood in places where it is most active. In a small experiment in which nonsense words were rhymed, both sexes used the left side of the brain to do the job, but some (but not all) of the females used the right as well. For dyslexics, too, brain scans show that boys make less use of the left half, the main site of the language centre. Perhaps that helps to explain why they suffer more from the condition than do their sisters. All these claims rest on small samples and all need more research.

Sex may influence the contents of the cranium, but it gets a bad press as it does so. The masculine mind responds through a long chain of internal and external events to its potent little switch, the *SRY* gene. Some of what it produces is less than attractive. Crime, for example, is a hobby of half the population. A man is ten times more liable to murder (and, to balance the equation, five times more liable to be a victim) than is his opposite number. All this is not just a product of civilisation, as in a certain South American tribe almost half the men have killed another while their partners almost never become aggressive.

The notion of a physical difference between criminals and others goes back to the days of phrenology. Most of the assertions are fantasy, but some geneticists claim to have come up with an explanation of at least a few cases of violent crime.

The film *Alien 3* is set in a planetary penal colony. Those condemned to it have inherited an extra dose of masculinity (or at least an extra Y chromosome), a state shared by one boy in a thousand or so back on Earth. They are, says their jailer, 'All double Y chromos, all thieves, rapists, murderers, forgers, child molesters . . . all scum' (the scum have turned to religion, which calms them down a little). Al Gore in his

failed fight against the reactionaries also referred to his oppo-
nents as the extra-chromosome right wing (which one he
meant is clear from the context).

The claim that an extra measure of manhood turns its
bearers into supermen – be they criminals or conservatives –
is simple, attractive and, for the most part, spurious. It began
with a study in Scotland in the 1960s, in which the inmates
of an institution for violent offenders had an incidence of
extra Y chromosomes twenty times higher than in the general
population. At once the chorus rose: such people were born
evil.

With a mere two XYY cases in the prison, the result may
well have been a statistical fluke. After all, there are some
twenty-five thousand such men in Britain, the vast majority
of whom have no idea of their condition and never get into
trouble. Without doubt, some boys with an extra Y develop
less rapidly than usual, talk later and are more active than
their fellows. They often end up two or three inches taller
than average and, with a slightly lowered IQ, may move to
manual rather than academic jobs. Almost all have no trouble
with the law.

A larger Scottish survey uncovered sixteen XYY cases
among the thirty-five thousand tested. Between them, they
had thirty convictions – more than average – but all were
minor and half were committed by a single XYY Scot who
had become a career criminal. Research in Europe has found
no detectable effect of the added chromosome.

For XYY, what seemed a simple story has become more
ambiguous as the information grows. Whatever the effects of
an added Y (and an extra chromosome is a substantial burden)
the idea that an additional dose of manhood rules the lives
of its bearers is simply wrong. Crime may be a masculine
pastime, but it is restricted to two segments of the male
population, neither defined by genes, the young and the poor.
Murder in the United States is twenty times more frequent

than in Japan and in fourteenth-century England was far commoner than today – but everywhere a man is many times more liable to kill than is a woman. The chromosomes are identical in each place and at each time, and the behaviour reflects society more than DNA.

The tale of testosterone – the epitome of manhood, from leadership to murder – is even more muddied. Without doubt, the substance influences the mind as it moulds the body, but it has, like the extra chromosome, been much over-sold.

Those whose hormonal tank is empty do sometimes lose interest in erotic pastimes. Castrated males given testosterone often find a renewed interest in copulation, and behave in more masculine ways. A small dose also improves the overall mood of eunuchs who are depressed about their state. However, castrati are not consistent, as the gelded unfortunates once held in American mental asylums were noted for their calm temperaments, while the word itself comes from the Greek term for a keeper of the marriage bed (a job which calls for a certain belligerence). Xenophon, too, insisted on such warriors as his guards, for, he said, they were the fiercest of all.

At the other end of the scale, men who brim over with unnatural amounts of the substance may also deviate from the norm. A million Americans abuse anabolic steroids, with a black market worth a billion dollars a year. Some take absurd quantities and a few die as a result. In one study, a quarter of the abusers who met a premature end were murdered, a third committed suicide and a third were killed in accidents. Some victims used other narcotics as well (and most people who misuse the drugs stay alive), but their fate hints at the chemical's power.

The role of testosterone in day-to-day life is harder to study because it is so labile. Even a good meal causes the levels to creep up. Emotion, too, plays a part, as it shoots up in footballers and chess players if they win, and plummets if they lose. The same is true of the fans.

Americans love to test each other. Thousands have been screened for testosterone levels and a blizzard of scientific papers hails chemistry as the engine of the nation's behaviour. Its young men have, they say, uniquely high levels of the substance. Those found guilty of violent rather than non-violent crimes and those who cause trouble in prison have the most (even if both remain within the range of citizens who stay within the bounds of the law). Men who beat their wives have more than husbands who leave their spouses unharmed. The divorced have more than those in stable marriages, and – so the story goes – war veterans with lots of it drink and drug themselves to an unusual degree. Even the male lawyers who defend them appear to have higher levels than do the lacklustre race who deal in commercial matters. The claim that bass singers have more of the virile essence and ejaculate more often than usual is disputed by baritones (who do, after all, get to sing the part of Don Giovanni).

Few of the assertions about hormones and behaviour stand up on a closer look. Scientists find it easier to publish positive than negative results and many of the latter no doubt remain unreported. Upper-class individuals with lots of the substance are less violent than their blue-collar equivalents, and the notion of a simple tie between hormones and dominance is quite wrong. Builders have more than architects, but architects are their bosses. The ambiguity of cause and effect remains. Do aggressive men have a different biochemistry, or do lust and rage change the balance? Attempts to define a 'testosterone syndrome' founder on the inability to predict the behaviour of those who are supposed to suffer from it. Nobody has ever identified a potential criminal from his steroid levels.

Homosexuality was once seen as a side-effect of that protean condition; the most unmanly of fates blamed on a shortage of the essence of manhood. Eugen Steinach, the inventor of the operation used by Yeats to restore his virility, had the cure: remove a testicle and replace it with another, either donated for medical reasons or a gift from an executed prisoner. Ten thousand people were treated, and the operation was seen, by those who performed it, as a great success. It lasted until the 1950s (almost as long as the alternative therapy, which blamed an excess rather than a deficiency of male hormone and used oestrogen to combat such behaviour).

The Bible condemns a taste for one's own kind ('Thou shalt not lie with mankind as with womankind: it is an abomination . . . both should be put to death') and the Catholic Church still refers to such acts as an intrinsic moral evil. Some priests even refuse to give the last rites at the death beds of AIDS victims, unless they repent.

Homosexual attraction is ancient indeed. The corpse of Ötzi the Iceman was claimed by a Viennese gay magazine to contain sperm in its anal passage, but the tale turned out to be fantasy (although Siberian images of the time do show a man on skis attempting penetration of an elk). According to Kinsey, a third of American men had experienced it, and a tenth spent long periods in the sole pursuit of males. His own interests (he went in for orgies with his colleagues) may have coloured his statistics, as today's estimates are that just one person in thirty to fifty lives as a more or less exclusive homosexual. In Britain today one man in six admits to at least a brief encounter with a partner of the same sex.

Plenty of animals go in for the pastime. Darwin himself noted that male monkeys present their rears to others and are mounted by them (whether mounter or mountee is dominant is not certain). Chimpanzees often use an erection as a display to their rivals and baboons grab a friend's testicles in greeting. Some male sheep, oddly enough, show no interest of any kind in females and confine themselves to the courtship of rams.

Such habits are often blamed on (or justified by) biology, but the evidence is as equivocal as is that for testicle transplants.

The term 'homosexual' was invented in 1869 by a German legal reformer. In the fine tradition that to name is to explain, those involved also became known as uranists or inverts, and their behaviour was classified into forty-six distinct types, each with its own symptoms. Medicine found it hard to make up its mind whether the condition was an illness at all. In the 1950s the *Diagnosis and Statistical Manual of the American Psychiatric Association* listed it as a 'sociopathic personality disturbance', a term later weakened to 'sexual deviance'. In 1973 it disappeared from the list of recognised disorders.

Homosexuals attract many theories, but fewer facts. Some psychiatrists blame an imperfect relationship with a father, and note that men with the preference tend to have lots of older brothers. They are also rather underweight at birth, perhaps because their mother was stressed in pregnancy. A certain small part of the brain may contain more nerve cells in men than in women. Gay males, some say, have intermediate numbers, but the evidence is weak at best.

Havelock Ellis (he of the upright female urinators) saw the habit as an inborn fault no more serious than colour-blindness. The idea of a gay gene re-emerged in the 1990s. It was hailed by much of the community as an alibi because, if the preference finds its roots in DNA, then it is at least not catching and should not be used to condemn those who practise it. The *Daily Mail*, in contrast, greeted the claim with the vile headline 'Abortion Hope after Gay Gene Finding'. Such preferences might indeed have an inborn foundation, but as usual when it comes to genes and human nature, reliable evidence is hard to find.

In one survey identical twins – who share all their genes – were twice as liable to share a same-sex preference as were non-identicals (who hold half their heritage in common). Such twins are similar for reasons other than DNA, as they often share a difficult life before birth and may copy each other as they grow. In addition, the sample was recruited through a gay magazine and men whose brothers do share their homoerotic interests may have been readier than average to come forward (which would place undue emphasis on any cases that might to some extent depend on inheritance). In any event, the result did not stand up in a later study.

A hint of an actual gene came from an alleged tendency for gay men to have gay relatives on their mother's (but not on their father's) side. Such a pattern is expected if the condition is influenced by a gene on the X chromosome (which, for sons, comes from the maternal, rather than the paternal, line). It was once tracked down to a small part of the X, but its very existence is mired in claim and counterclaim (which, to be fair, is true of most genes supposed to influence behaviour). A second study could not find the supposed gene at all, and the original work used, as had the twin research, families with more than one gay member. Perhaps, as for schizophrenia, depression and other rare patterns of behaviour, the preference has a genetic component in some people but not in others. A partiality for the same sex is, after all, just an extreme of the normal range of activity. Nobody claims that single genes are responsible for tall or clever people and there is no more reason to do so when it comes to our diverse erotic interests.

Doubts about the existence of gay genes have not stopped guesswork about their evolution. Perhaps they boost the erotic activity of a man's sisters, and so help to pass themselves on. Homosexuals might help relatives with childcare and push up their own biological chances. Such notions are easy to invent but harder to test. Surveys show that such men are in fact no fonder of their families than anybody else. They live further from their parents and siblings than average, and are less generous when they get into trouble. The finding does not discourage enthusiasts for biological explanation: in ancient times homosexuals were, no doubt, crucial for the survival of their sisters, their cousins or their aunts.

Whatever the truth about genes, sexual preference, like crime, intelligence or height, is much related to circumstances. *Tom Brown's Schooldays*, published in 1857, speaks of 'miserable little pretty white-handed curly-headed boys, petted and pampered by some of the big fellows, who . . . did all they could to spoil them for everything in this world and the next' (with an admonitory footnote that 'There were many noble friendships between big and little boys, but I can't strike out the passage: many boys will know why it has been left in'). Homosexuality at Rugby School (upon which the book is modelled) flourished in the absence of other outlets and most pupils abandoned it as soon as they had a chance. The same is true today. People confined in prisons may turn to the activity even if they would otherwise avoid it, but some who live in freedom choose homosexuality even when the other sex is present.

Genetics is a science of the extremes. It does well with rare illnesses such as cystic fibrosis, but the inheritance of common ailments such as diabetes and heart disease is harder to pin down. For variation in height, weight or personality the situation is even more complex. DNA is without doubt in part to blame, but the more details that emerge the more complicated the story becomes. The genes involved differ from place to place, and from family to family. Some diabetics and depressives suffer for genetic reasons, while others are ill because of the way they live or because their constitution cannot deal with what the environment offers them. To separate nature from nurture is difficult, and may not even mean much.

Male behaviour is much the same. For the norms, rather than the extremes, the importance of genes fades in the face of the environment, and biology has rather little to say. When it comes to understanding the conduct and character of men when compared to women, vulgar supposition is easier than objective measurement. Manhood tells a social tale as much as one written in nucleic acids. It must, with all that it implies, be constructed, and once its foundations are laid, what rises from them has little to do with DNA.

THE DESCENT OF MEN

The final pages of *The Descent of Man*, the model for the present work, contain a lofty account of the natural superiority of men over women: an ascendancy driven, thought Darwin, by sexual selection. In the nineteenth century such ideas seemed self-evident. Now things have changed.

Males, towards the end of the last millennium, felt a sudden tightening of the bowels with the news that their services had at last been dispensed with. Dolly the sheep – conceived without masculine assistance – had arrived. Her birth reminded half the population of its precarious position. Perhaps, some fear, science will cause nature to return to its original and feminine state and men themselves to fade from view. The Y chromosome, after all, is a mere remnant of a once mighty structure, which might in a few million years disappear. Why should those who bear it feel any more secure?

They have good reason to worry. Even without the help of the cloners, males are wilting away. To be a man has always been a minority interest but the cruel facts of life and death have made his position even more precarious. From sperm count to social status, and from fertilisation to death, as civilisation advances those who bear Y chromosomes are in relative decline. Even their place in society – once undisputed – is challenged by the advance of womankind. The flames of

Prometheus, the bringer of fire, have been much dimmed since Darwin's day.

Not all the news is bad. The world has four hundred billionaires. Nine out of every ten are Prometheans, half of them self-made (and just one of the women, the founder of the Gap fashion chain, earned rather than inherited her money), but be they plutocrats or paupers, all men pay a high price for their privileges. What once seemed a natural superiority has been lost in the face of a manifold failure to deal with modern times.

Homo sapiens, in the certainty of his dominion, has changed the world. Much is for the better, and both sexes now live longer, healthier and perhaps happier lives than before. Infant mortality in much of the Third World is down to what it was in the West in 1970 and global life expectancy is up by half a decade. For most people, existence is easier than it has ever been, but the minority blessed with a sex-determining gene has not had its fair share. When the going gets tough the women do better, and when life is good men indulge so much in its vices that they suffer again.

In the nineteenth century, thousands of Mormons travelled west to Salt Lake City. The poorest were forced to push their goods on handcarts, five to a cart. In July 1856 the Willie Party, four hundred strong, set out on the thousand-mile trek across the mountains. With masculine lack of foresight, they started too late and, in the Rockies, winter caught up with them. As a survivor wrote, 'At first the deaths occurred slowly and irregularly, but in a few days at more frequent intervals, until we soon thought it unusual to leave a camp-ground without burying one or more persons . . . Men who were, so to speak, as strong as lions when we started our journey, and who had been our best supports, were compelled to succumb to the grim monster . . . Many a father pulled his cart, with his little children on it, until the day preceding his death. I have seen some pull their carts in the morning, give out during the day, and die before next morning.'

Sixty-eight of the emigrants died. For those over forty, ten times as many men succumbed as did their partners, and for the young the rate was twofold. Those with the strength of lions were, when it came to the test, the weaker sex.

For people blessed with a Y, life is hard from the start and gets harder. Fertilisation produces a slight excess of male embryos (perhaps because the relevant sperm swim faster) but from then on things go downhill. Even before birth, males cope less well. There are, as a result, more girl twins as boys do not survive the tough conditions in a shared womb. As the foetus grows, boy babies suffer more brain damage and more birth defects. In addition, they experience a whole host of diseases, from colour-blindness to muscular dystrophy, caused by the damaged genes found on the X which are usually hidden in girls but always exposed in their brothers. At birth they are in a mere 5 per cent excess over their sisters and not for a month does a boy child reach the level of maturity of a newborn girl.

Then the rot sets in. Up to the age of fifty or so, the balance is about even, but at eighty a mere third of the population consists of men, and the Queen sends nine times as many congratulatory telegrams to her lady centenarians than to those of the opposite sex. From middle age onwards it is a woman's world.

Such a large gap between husbands and wives is new (although Swedish records from the eighteenth century, when life expectancy was half that of today, show that their women already did a little better than the others). A century ago, American husbands and wives died at about the same age. Even at the time of Pearl Harbor, men and women who made it to retirement had just over a decade in which to relax. Now newborn American girls can expect an eight-year advantage and in England the average male lasts to seventy-three, while his partner has six years more to go. Some of the shift comes from a gain by one group because of healthier child-birth (in Victorian times the commonest cause of premature death, but now with a tiny rate of mortality) but more comes from man's own inborn frailties and his stubborn refusal to recognise them.

Why do people with the fatal gene do so badly? Society plays a part, and stupidity helps, but the testes are much to blame. Some of the harm is direct, for their secretions damage those who bear them. The cells of the immune system die when exposed to testosterone, and the male body, with its vast quantities of the substance, is less able to make antibodies than is its opposite number. As a result men find it more diffi-cult to fight cancer and to resist infection by a variety of diseases, from gonorrhea to rabies. In most animals males die younger than their opposites (by thirty years in sperm whales), as a hint of the importance of chemistry both in its direct effects and as a promoter of rage and lust.

Testosterone is a signpost to several well-trodden paths to oblivion. Suicide is now the main cause of death among young men and is three times more dangerous to sons than to daugh-ters. The incidence among boys has doubled since the 1970s, while in girls the figure has stayed much the same. Other violent deaths, from accident or murder, are on the way down but – once more – males have been slower to gain and boys still face a doubled risk of accidental death compared to girls. The difference is large even in four-year-olds.

By 2010 a third of all British males will live alone. For bachelors (but not spinsters) mortality is twice that in the population as a whole. Darwin noted as much, and *The Descent of Man* gives several paragraphs to the doubled death rate among unmarried, compared to married, Scotsmen of his day (he used the figures to promote the eugenical agenda that the feeblest stay single). Men, it seems, evolved for matrimony, as they gain several more years from the state than do their partners. 'Griefe' was entered as a cause of death in the rolls of mortality in the seventeenth century and widowers still do far worse than widows when forced to deal with their un-welcome situation. Most of them meet their ends from cardiac problems. They die, literally, of a broken heart.

Misery and violence kill off plenty of males, but neglect destroys far more. Men do not like to bother their doctors, or even themselves. As long ago as 1662, John Gaunt, the founder of demography, noted that physicians had two female patients for each one of the opposite sex. The latter, he thought, died of their vices while women fell victim to the infirmities of their state. He was not far wrong, for females under forty still go to the doctor twice as often as do their husbands. Part of the difference comes from the troubles of birth control and pregnancy, but most does not – and that is odd, for males have far worse health.

Each year in Britain, five million man-years of life are lost to a demise that could, in principle, be avoided. Smoking, fat and stress all specialise in one segment of society. From gout to hernias (four and nine times more common in males) men suffer while their partners are spared.

One man in every three smokes, one has given up, and one has never had a cigarette. Those in their twenties indulge in the murderous pastime most of all. Girls have been lured into the habit by tobacco companies (and lung cancer has over-taken breast cancer as their commonest malignancy) but so far have not been forced to face the full consequences. When it comes to alcohol, the difference is starker. With a standard unit set at half a pint of beer or a small glass of wine, the (some-what meagre) recommended limit is twenty-eight units a week – which is around two pints of beer, or half a bottle of wine, a day. Twenty-year-old youths *average* twenty units a week, and half imbibe more than the safe limit. Self-delusion extends to the pub and a respectable minority of dangerous drinkers believe that they drink either 'a little' or 'hardly at all'. Those who soak up more than the recomended amount die at twice the rate of those who imbibe very little. A small dose has a minor – and much-publicised – beneficial effect, but alcohol is a killer. Drink causes cancer of the pancreas and throat and damages the liver, and is associated with the majority of murders and of deaths in fires.

The demon in the bottle also acts as a chaperone to the biggest assassin of all. Diseases of the heart and circulation account for much of the shortage of men. They cause half of all male deaths before retirement and the overweight, tobacco-ridden and stressed inhabitants of a British public house are in particular danger.

Heart attacks are rather new. From the First World War to the Second the overall incidence doubled, and from then until 1980, doubled again in men – but dropped among women. Over the past couple of decades, males have done a little better, but less so than their partners, and they now face three times their risk. Young men do even worse, with seven times the rate compared to women of their own age. Oestrogen protects the heart, as it increases the release of nitric oxide and relaxes its blood vessels. Those with Y chromosomes lack that shield, and behave in such a way as to damage themselves still further. The epidemic of cardiac disease affects a small global minority: men in developed countries. The gender difference is greatest in places with a fatty diet, which allows Scottish males to boast of – among their other undoubted achievements – more heart attacks (and fewer teeth) than anybody else on earth.

In Glasgow's vast Victorian cemetery, the Necropolis (dominated by a statue of the reformer John Knox, but now the haunt of alcoholics and addicts about to fulfil their own destiny), the size of each gravestone, from marker to massive stele, tells the tale of sex, wealth and death. A century ago people died younger than today, but the effects of class and gender were much the same. The height of each monument (and some tower over the visitor) reflects the circumstances of the person beneath; the richer he was, the bigger the stone and, the epitaph shows, the longer he lived.

Today's markers are feeble in comparison, but the tie between manhood, mortality and money is as strong as it was in Victoria's day. In 1974 Edward Heath's government was defeated by the strikes of – among others – the Scottish miners, but by 1997 the revenge of his Tory successor had rendered their society more or less extinct. In those years the gap between rich and poor became wider, in Scotland as elsewhere. About half of all Scottish fifty-five-year-olds alive at the period's beginning had died at its end. More men died than women, of course, but the effects of class on each sex were strong. The least well-off Scottish men had the pattern of death of their richer brothers seven years older than themselves. Among the poorest fifth, three-quarters died, while only half the richest men met their end. For women the effects of wealth on death were far less marked.

In modern Britain, after the political revolution ushered in by New Labour, a baby born in the least affluent place (in Bootle, in Liverpool) can expect to live for a decade less than one born in the richest, in south-east England. Even within a single city the contrasts are stark: six stops on the Jubilee line, London Underground's link between Westminster and Newham in the East End, take six years off a newborn's life expectancy; a year for each tube station. Everywhere the forfeit on poor boys is greater than on their sisters and although the gap in survival between rich and poor is closing for women by a month for each year that passes, it shows no sign of so doing for men. Those who carry the *SRY* gene find it harder, of their nature, to deal with poor nurture.

Politicians, of whatever flavour, worry about 'social capital', the kind neighbours and safe streets that help make communities succeed. A low score is associated – quite apart from the effects of tobacco, drink, age and poverty – with high blood pressure, depression and other illnesses; and the men who live in such places suffer more than do their wives. Their lives are made joyless by the foul behaviour of a few young people, whose habits have been medicalised into an illness called 'conduct disorder', which starts with tantrums, lies and theft and soon moves on. Twice as many boys as girls are so diagnosed, and half are later convicted of a crime. In Lambeth, in south London, their costs to society over two decades, in terms of health care, prison and the damage they cause, average seventy-eight thousand pounds a head for boys, and a tenth as much for their opposite numbers.

The United States faces a larger social problem. Young black men are, so conventional wisdom has it, maleness personified, good at sport, at sex and at violence. They suffer for their stereotype. They are murdered at a rate equal to that of all whites and all black women put together. Black Americans with a college education have about the same number of years of healthy life as do whites (a proof that biology has nothing to do with it), but their fellows at the bottom of the heap have a sixteen-year reduction in life expectancy. Heart disease and cancer play a part, but violence is almost as important – and, once again, black women do much better, with a far greater mortality gap for black husbands and wives in late middle age than for whites.

Men, such figures show, are worse at dealing with stress than are their opposite numbers, but to a biologist this scarcely matters, for most of them have had their children and fulfilled their evolutionary destiny before they meet their miserable ends. Evolution cares only about genes: about inherited differences in the ability to reproduce. Now, some say, stress threatens man's very reason for existence. It has moved from the streets to the bedroom.

Each generation of Americans since the Civil War has (with the exception of the post-war baby booms) had fewer children than the one before – and, in 1990, for the first time, households without children outnumbered those with. The whole Western world faces the same problem. The press is full of rumours about a collapse in virility. Perhaps, as men suffer the pains of modern life, so do their vital parts.

In 1950 Americans known to be fertile because they already had a child had an average count of around a hundred million sperm per millilitre. In 1974 came a shock. The numbers had dropped by half, and a mere one husband in twenty could reach the previous standard. At once a series of surveys began – with mixed results. They show a reduction in the Scottish count each year from the 1950s to the 1970s, a decrease in the ability of their English equivalents to swim and an increase in the proportion of damaged sperm in Denmark. An already well-endowed Finland, in contrast, showed a real improvement in both the number and the quality of such cells. A recent worldwide study of thousands of volunteers suggested a drop from over a hundred million sperm per millilitre of semen in 1940 to just sixty-six million half a century later. At this rate, Westerners will make no sperm at all by the end of this century and the population crisis will be solved.

Global sperm failure is alarming but ambiguous. Vigour varies in a mysterious way from place to place. New Yorkers manage a hundred and thirty million cells per millilitre, rural Minnesotans do no better than a hundred million, while hedonistic California lags behind with seventy-three million. Finns match the men of New York, while Parisians and the people of Edinburgh (otherwise not much alike) fall well behind.

Such wild swings hint at problems with the figures rather than the donors. Some come from infertility clinics and some from subjects who already have a child. Some centres collect in the morning (when semen quality is lower) and others in the afternoon. A few insist on abstinence before the sample, while others are more generous. Husbands produce fewer cells when they spend time with their mate, which also confuses the issue. All this can cause a tenfold difference in any individual's output from day to day. In addition, technology has so much improved that the older measures (and in particular the crucial high counts from the 1940s) could be overestimates.

A new survey of potency has now begun in four European cities – Edinburgh, Paris, Copenhagen and Turku in Finland. It involves men under forty whose partners are pregnant, with a real attempt to standardise the conditions in which the data are gathered (some problems remain; the average Dane manages a mere three days without ejaculation before an appointment, while his abstemious French equivalent can hold out for almost a week). The lowest counts were in Copenhagen, followed in turn by Paris, Edinburgh and Turku (which came a clear top). The citizens of Edinburgh should be proud of their cells' ability to swim, which takes the European gold medal.

The work goes on and may reveal changes in male quality with time. Whatever it shows, the unsettling fact remains that such counts do vary in parallel with other reproductive problems. Undescended testicles, prostate cancer and abnormal penises are all much more frequent than they were and, in the West, the incidence of cancer of the testis has doubled in twenty years. Danes have the lowest sperm counts in Europe, and one Dane in a hundred will get the disease. Testis cancer starts long before birth and results from an error in the cell line that leads to sperm themselves. Perhaps the global crisis is real, and all these troubles are symptoms of a new malaise.

It may come from a joint attack upon male biology by the two great engines of the modern world, physics and chemistry. Sperm are made best at two or three degrees below body temperature, which is why the testes find themselves in such an exposed position. Even in frigid Edinburgh fewer emerge in summer than in winter. The first hint of the dangers of modern life came from Italian taxi drivers. In their celibate land (in 1998 half as many Italians were born as forty years earlier, with an average family size of just over one child), cab-drivers have a high proportion of damaged sperm and find it harder to conceive than average. A cramped seat is at the base of their problem.

Because a driver's legs are kept together, body heat reaches certain parts that evolved to avoid it. His most delicate objects are chilled by an ingenious mechanism. The warm arterial blood on the way out passes heat to the colder blood as it returns, to keep the reproductive department cool. In the driver's seat, the network is short-circuited and after a couple of hours behind the wheel the scrotum is two degrees warmer than after the same time on foot. All drivers suffer, and men who are wheelchair-bound are at particular risk as their testes never get a chance to cool down.

There may even be a fit between a lowered sperm count and time spent on the steamy London Underground. A sauna does the job even better (and it can take weeks to get back to normal). Even worse, boy babies who wear plastic nappies are warmer below the waist than are those in cotton. Nappies themselves go on for longer in the lax Western world. In East Germany most boys were toilet-trained by the time they were twelve months old and could put aside their childish wear in favour of short trousers, while in the West the ritual − and the drop in testis temperature which follows − is delayed for two more years. Boys whose testicles are hot because they do not descend from the body cavity tend to be infertile, and wearers of high-technology baby wear may pay the penalty when they grow up. One treatment uses a fan directed at the organs while their owner sleeps. Three months of determined ventilation pushes the numbers of sperm up in a hopeful way and may reverse the malign effects of modern infancy.

American homes contain a hundred thousand artificial substances in plastics, paints, pesticides, dry-cleaning fluids, detergents and more. Most are harmless, but some have sexual preferences of their own.

Chemists are, in general, male, but what they make has feminised the world. Forty years ago, hermaphrodite fish began to appear in European lakes and streams. In the River Aire in Yorkshire, every male fish showed signs of a desire to change his identity, with similar effects downstream from sewage works on almost all other rivers in the region. Then a plague of ambiguity spread across the United States. Alligators in Lake Apopka in Florida suffered from small penises; female seagulls took up with each other; and what could be discerned about the sperm count of panthers showed a slight drop. In Japan, too, the males of certain marine fish began to make eggs as well as sperm. In Britain's seas, in contrast, female dog-whelks became endowed with phalluses.

The whelks were confused by a new tin-based anti-fouling paint that interfered with development. The average yacht spends a day a month at sea and passes the rest at anchor, leaking poisons from its bottom. The guilty chemical was easy to track down, and has been banned for use on pleasure craft. The whelks have recovered, but men may find it harder to do so.

Few chemicals confer maleness, but many take it away. Which, if any, are responsible for our own troubles is hard to say.

The Pill changed men's lives in more ways than one. It caused reproductive hormones to leak into tap water and has been blamed both for the sex changes in freshwater fish and for the drop in our own sperm count. The jury is still out on the issue, but other hormones have had a disastrous effect. A drug called diethylstilbestrol was once thought – in error – to prevent miscarriage. Five million mothers took it and for a time it was even used as a chicken food supplement. A third of the boys exposed to the drug in the womb suffer from small testes or a reduced penis. In rats, the chemical causes prostate and testicular cancer (although there is as yet no sign of those problems in ourselves).

To give a powerful steroid to pregnant women was at best unwise, but the effects of other chemicals were harder to foresee. The 1950s saw a wonderful new chemical treatment for banana pests. Soon the substance was much used. Twenty years later the workers noticed something odd: they had almost no children. Their sperm count had dropped by five hundred times.

Many other substances mimic hormones and bind to the cellular receptors for the body's internal messengers. Dozens have been accused, often on weak and circumstantial evidence, of interfering with man's chemistry. They include the break-down products of household detergents, the pesticide dieldrin (whose molecule is shaped rather like oestrogen) and the plastics used on the inside of some tin cans. Spermicides, the film used to seal dental fillings and polychlorinated biphenyls from the electrical industry are also on the list. In the sea, the millions of gallons of waste water pumped from oil wells contain natural chemicals that act to feminise fish. Even licorice, which might seem a modest vice, is a risk as it contains a substance which blocks certain enzymes in the hormone cycle. A week's indulgence causes the levels of the virile essence, and the libido itself, to sag.

Milk might also play a part, for the product of the modern cow – which often comes from animals who lactate while pregnant because they are given large amounts of high-protein food – has lots of oestrogen. Japan, where much more milk is drunk than in earlier times, has shown a faster rate of increase in prostate cancer than anywhere else (and Denmark, a nation of dairy herds, shares the problem). Beef, too, may not be blameless. American farmers use some of the most potent oestrogen-like substances ever made as growth promoters. They are banned in Europe (which has led to a trade dispute between the two continents) but in the United States are almost unregulated. They may leach into the water supply around the pens in which cattle are fattened and might persist in meat on sale. Little is known of the long-term effects, if any, of a diet with added super-hormones. For milk-drinkers and carnivores, a genital disaster might be about to happen.

Those who avoid meat may feel smug about the problem (as we do about many things) but have worries of their own. Natural does not always mean nice. In Australia a million sheep a year do not produce lambs because they eat too much clover, which contains an oestrogen-like material. Keen vegetarians are fond of soybeans – which contain lots of such substances. Elderly Japanese-Americans who eat tofu (an essence of soybean) may lose their mental powers earlier than others, perhaps because the compound interferes with oestrogen, which protects the brain.

As factory owners point out, nobody – with the exception of banana workers – who uses, or makes, such products has suffered a drop in sperm count. Organic farmers in Denmark have the same number of male cells as do those who use pesticides, and workers in plastics factories do not deviate from the norm. The amounts of alien substances in the normal diet are tiny. Even so, few people have been exposed to large quantities and it may be hard to pick up a small general effect. Thousands of animals have been treated in the laboratory – albeit with far more of the substances than any person is exposed to – and some have been femi-nised. They can accumulate in body fat, and might work together so that a cocktail of several unrelated compounds has the same effect as large doses of a single one. The case against chemistry is far from closed and the European Commission has set up a task force to examine a hit list of more than five hundred suspects.

Whatever the truth about feminisation, manhood itself is in full retreat. The crisis comes not from outside, but from within: from the great loss of self-confidence that has swept across half the world. Adlai Stevenson, who advised the 1955 graduates from Smith College (a women's university in Massachusetts) to mould their lives around an attempt to 'influence man and boy . . . through the humble role of house-wife', would be booed off stage today. Women who read magazines such as *Cosmopolitan* in those days were regaled, as now, with pictures of models. About a fifth of the females were undressed to some extent, but – given the values of the time – just one in thirty of the male models had shed even a single garment. Nowadays, twice as many men as women display their unclothed flesh for the delight of the lady reader.

The new economy exposes man's weak points in other ways. From his point of view, the more equal society of the twenty-first century is not an advance, but a retreat to a long-lost feminine Utopia.

Myths of matriarchy extend well beyond the Amazon, and several tribal peoples have legends of a sexually equitable past. Almost all describe a moment when the child-bearers lost their power because of a moral failure that forced the other half to take charge. To Friedrich Engels, in contrast, woman's loss of power came from economics. The 'world historical defeat of the female sex' coincided with the origin of agriculture, a time when landowners forced submission – and monogamy – onto their inferiors. Poverty-stricken as a peasant might be, his wife was even poorer. She owned too few resources to make any kind of life and had no choice but to play a secondary role.

Now the world has taken a great leap backwards – out of the fields and into the supermarket. For most people, life has become more like that of a hunter-gatherer. The return to an earlier pattern of existence has, in part, avenged woman's historical defeat.

Western civilisation has moved away from the peasants who prevailed for ten thousand years. People no longer grow their own food, but sally out to work and to shop and return home to eat and to take it easy. Goods are not made by each family, but are foraged for in the wider world. Unlike farmers, almost nobody stays where they were born, and the importance of property in decisions about whom to marry has faded away.

Hunter-gatherers spent far less of their time at work than do peasants. Their lives were full of leisure as, once the search for meat or seeds was over, the day was done. In southern Africa, the few !Kung bushmen who hold out in part to the older way of life have to work for just two days a week to feed their families, a fraction of the time passed by the labourers in the nearby diamond mines. Their wives are much more independent than are those of the local farmers. In today's West African peasant societies, too, women work three hours a day longer than their opposites and have much less time to themselves. Their ancestors in the new economy of ten millennia ago had an even more miserable existence. Women's bones reveal them to have been stooped and ill-nourished, with damaged joints caused by a life as grinders of grain.

The twentieth century's giant step into the past has liberated the rich more than the poor, and women more than men. In 1950s Britain the average man worked for forty-seven hours a week, for forty-seven weeks a year, for forty-seven years, to give a hundred thousand hours of lifetime labour. Now those figures have been halved. For the middle classes, with their new ability to gather what they need, leisure has returned. It has shifted the position of the sexes. Women have more work to do outside the house than once they did (in a 1940s poll most Americans felt that they had an easier time than men did but now the figure is reversed), but they gain much more from their labours.

Many men welcome women's liberation, but it means, too, often, a decline in their own importance. The changes in today's society show how masculinity is constructed by property as much as by genes. The economic tide has begun to turn, at man's expense. With the decline of industry, that male preserve, employment has shifted to services. Women, many employers feel, do such work better (and cheaper) than their partners. American blue-collar men have become poorer in real terms since the 1970s, but their working wives have gained. In Britain, too, at the depths of the 1980s slump more than half of the employees in the once great ship-building cities of Tyneside were women. The proletariat, in decline as it is (in the 2001 census, more than half of all British citizens identified themselves as 'middle class' or 'classless'), has less need for males.

The figures are stark. At the end of the Second World War, husbands were in effective control of all family finances. As recently as the 1970s, British wives could not obtain credit without their approval (and in Spain until the end of the decade they needed a partner's permission to work at all). Twenty years later, independent taxation for couples was introduced. Now three-quarters of all married women have a job. In the 1960s they earned half what their husbands did but now the gap is far less (and, for people under thirty, scarcely exists). They may not overtake their opposites in income terms for a number of years, but their relative ascent is dramatic.

Until two decades ago, girls were the deprived group in other ways. In the Third World they had half the chance of education as did their brothers. Now the gap has almost gone. Certain countries still give boys their due, but in many emergent nations – South Africa included – more girls than boys go to school. In the West, too, the second sex has raced ahead. The United States has two million fewer males than females in universities and, in Europe, more women graduate than do their opposite numbers. At this rate, in ten years or so the proportion of educated men will be half that of women and, quite soon, more women than men will have jobs. The poor, needless to say, do worst of all, and among American blue-collar families, university-educated daughters outnumber sons by threefold. Jane Fonda, who has given millions to the Center for Gender and Education at Harvard, might be alarmed to see most of the cash directed to a group other than her own.

In a world filled with professions rather than trades, macho values have become an embarrassment. The managers of today prefer adaptability and intuition to the aggression of days gone by. The present century may be the age of women; the first in which, like it or not, slightly less than half the population is forced to accept that biology no longer gives it an alibi for injustice.

To optimists, all this is progress, a fulfilment of Marx's prediction that 'The immediate, natural, necessary relation of human to human is the behaviour of man to woman . . . From this behaviour one can judge the whole stage of human development.' Perhaps we have at last developed to the stage when men treat the rest of the populace as fellow humans, and no less (or more).

Such crises of masculinity have happened before, and have been overcome. In Restoration England the grim Puritan was replaced by the fop ('The Men, they are grow full as Effeminate as the Women . . . they sit in monstrous long Periwigs, like so many owles in Ivy Bushes, and esteem themselves more upon the Reputation of being a Beau, than on the Substantial Qualifications, of Honour, Courage, Learning and Judgment'), but he in turn disappeared in favour of true manliness.

Two centuries later, the United States, too, began to worry that its males had lost their *raison d'être*. Young ladies took over as schoolteachers, to cries about their spinsterish attitudes and the effeminisation of their charges. No longer were the norms of society upheld: 'the boy in America is not being brought up to punch another boy's head; or to stand having his own punched in an healthy and proper manner'. Even the Boy Scouts of America – who date from that era – were established to validate masculinity and to restore the primitive past.

Edgar Rice Burroughs's *Tarzan,* published in 1912, sold

thirty-six million copies to a nation convinced that its men were not living up to their potential. President Roosevelt sounded the alarm: 'The nation that has trained itself to a cancer of unwarlike and isolated ease is bound, in the end, to go down before other nations which have not lost *the manly and adventurous virtues.*' The success of Robert Bly's *Iron John*, that best-seller of the 1990s, with its call for Zeus Energy, Masculine Grandeur, the moist, the swampish, the wild and the untamed (and its rather pricey Wild Man Weekends) suggests that the cycle of self-flagellation goes on. Lash themselves as they might, most men realise, almost a century after Roosevelt, that their attempts to live up to their past have led to social and political disasters which a world of mass destruction can no longer afford. Unwarlike ease now seems a more natural condition than does some inborn fate of primitive superiority.

Man in the general sense (as French dictionaries say: *Homme – terme générique qui embrasse la femme*) has of course descended from other primates, as Darwin was the first to prove. Science since his day shows how men in the narrower context are the progeny of their Y chromosomes; but the discovery means less than is often supposed.

Not long ago, the biological basis of their state seemed a truth as certain as that of our descent from apes but – as this book has, I hope, shown – genes must often defer to social reality. Darwin's *Descent of Man*, in its portion devoted to selection in relation to sex, develops his notions of competition and choice and shows how decoration and belligerence in the animal world emerge from a male's selfish interest in passing on his heritage. Darwin was quite overt about ourselves. One party had evolved to be the leader: 'Thus man has ultimately become superior to woman. It is, indeed, fortunate that the law of equal transmission of characters to both sexes prevails with mammals; otherwise it is probable that man would have become as superior in mental endowment to woman, as the peacock is in ornamental plumage to the peahen.'

Darwin's great work of 1871 showed that *Homo sapiens* had a pedigree of prodigious length, but not, as he put it, one of noble quality. His proof that we live in a world of change caused an upheaval in scientific thought so complete as nowadays to be accepted almost without reflection. His view of society was less generous. As a true Victorian, he saw the social order as set in stone. To him, as to his contemporaries, the roles of men and women were as immutable as those of rich and poor, or of savages as compared to the world's civilised races.

Charles Darwin, should he return, would be delighted by the confirmation of his ideas about the origin of our own species, but much less so by the fate of his views on the nature of men. In the brief century and a half since his great insight into the past the world has moved on. Gender differences have been consumed by social change. We are in the midst of an ascent of women matched with an equivalent descent of men.

That sex may, in part, be able to change its habits to escape the fate encoded on the Y chromosome. So far it shows little sign even of facing up to the problem. Death itself is undecided which option to assume. Mortality was female for Shakespeare's Antony ('I will be a bridegroom in my death, and run into't, as to a lover's bed'), but for Hamlet the final exit was, as a 'fell sergeant', male. *La Mort* is feminine, but *Der Tod* is not. St Francis's last words were 'Welcome, Sister Death', but the Grim Reaper himself wears trousers, most of the time.

The century of progress in biology since *The Descent of Man* has given death a penis, and men themselves the comfort of knowing quite why their organ is so dangerous. What they choose to do about the problem is quite another matter, which – like so many of man's affairs – has little to do with science.

BIBLIOGRAPHY

Everyone is interested in sex, and most people yield, at some time or another, to the temptation to put their thoughts on paper. As a result, publications on males go from the sublime to the scientific and well into the ridiculous. That poses a problem both for an author and for any reader who might wish to learn more about the subject.

The sublime (much of English literature included) is in the main passed over here. Its opposite – which includes a variety of strange works that mix pop psychology with ancient myth, weak research driven by preconceived notions, and great clouds of soft-headed sociology – has proved harder to avoid. I have tried to ignore it as much as I am able in favour of science itself, with its unique ability to make the fascinating dull.

The mountains of dismal stuff on men are, fortunately, balanced by some of the best of modern biology matched with medical and historical enquiry of equal excellence. Much of the material is technical and I have not been able to do justice to it all. It stretches my non-specialist's understanding to the limit and beyond – 'RACE-PCR and sequencing of genomic clones led to the discovery that SCRs undergo 5'-splicing, which may regulate receptor expression during spermatogenesis'; or the French philospher Lacan's account of the penis ('The erectile organ comes to symbolise the place of *jouissance*, not in itself, or even in its form of an image, but a part lacking in the desired image: that it is equivalent to the square root of minus 1 of the signification produced above, of the *jouissance* that it restores by the coefficient of its statement to the function of lack of signifier . . .'). In spite of such difficulties I have at least tried to deconstruct part of the scientific study of men.

I cite below a variety of books and papers which should be reasonably accessible to the non-specialist; and concentrate on material published within the past five years, a period of unprecedented advance in the biology of maleness.

There are several first-rate texts on human reproduction (and, simple as their internal machinery might appear in comparison to that of their partners, men get fair coverage in most of them).

M. H. Johnson and B. J. Everitt's *Essential Reproduction*, 5th edn (Oxford: Blackwell, 2000) gives a medical slant on the tale, while M. Potts and R. Short in their *Ever Since Adam and Eve: The Evolution of Human Sexuality* (Cambridge: Cambridge Univ. Press, 1999) present a more general account of the emergence of mankind from his sexual past. Historians often indulge in sex for light relief, and their many admirable treatments of the strange tale of reproduction include C. Pinto-Correia's *The Ovary of Eve: Egg and Sperm and Preformation* (Chicago: Univ. Chicago Press, 1997) and, for ancient historians, T. Taylor's *The Prehistory of Sex: Four Million Years of Human Sexual Culture* (London: Fourth Estate, 1996). Those who wish to understand where the American male has come from should read M. Kimmel's *Manhood in America: A Cultural History* (New York: Free Press, 1996).

Sex is odd in itself (how very odd it can become in the animal world is illustrated in Olivia Judson's oddly titled *Dr Tatiana's Sex Advice to All Creation* (London: Chatto, 2002)), and some of those who study it are odder. Readers of a gloomy cast of mind may wish to turn to L. Osborne's *The Poisoned Embrace: A Brief History of Sexual Pessimism* (London:

Bloomsbury, 1993) while the positively suicidal can sample K. S. Guthke's *The Gender of Death: A Cultural History in Art and Literature* (Cambridge: Cambridge Univ. Press, 2000). J. Gathorne-Hardy in his *Alfred C. Kinsey: Sex the Measure of all Things* (London: Pimlico, 1999) tells the bizarre tale of the first modern sexologist (who was also a wasp expert). His conservative image hid a life of erotic recklessness equal to that of the great society of decadents who, as artists, admitted their own degeneracy in a series of immensely readable works, all of which are avoided here.

The following books and papers keep their authors' secrets to themselves and deal only with the doings of their fellow men.

Chapter 1: Nature's Sole Mistake

D. Blum, *Sex on the Brain: The Biological Differences between Men and Women* (New York: Penguin Books, 1997).

B. Charlesworth and D. Charlesworth, The degeneration of Y chromosomes, *Phil. Trans. R. Soc. B*, 355 (2000): 1563–72.

A. K. Chippindale and W. R. Rice, Y chromosome polymorphism is a strong determinant of male fitness in *Drosophila melanogaster*, *Proc. Natl. Acad. Sci. USA*, 98 (2001): 5677–82.

H. Ellegren, Evolution of the avian sex chromosomes and their role in sex determination, *Trends Res. Evol. Ecol.*, 15 (2000): 188–92.

J. A. M. Graves, Human Y chromosomes, sex determination and spermatogenesis: a feminist view, *Biol. Reprod.*, 63 (2000): 667–76.

A. Hossain and G. F. Saunder, The human sex-determining gene SRY is a direct target of WT1, *J. Biol. Chem.*, 276 (2001): 16817–23.

T. Kuroda-Kawagachi *et al.*, The AZFc region of the Y chromosome features massive palindromes and uniform recurrent deletions in infertile men, *Nature Genetics*, 29 (2001): 279–86.

B. T. Lahn, N. M. Pearson and K. Jegalian, The human Y

chromosome in the light of evolution, *Nature Reviews Genetics*, 2 (2001): 207–16.

U. Mittwoch, Genetics of sex determination: exceptions that prove the rule, *Mol. Genet. Metab.*, 71 (2000): 405–10.

I. Nanda *et al.*, Three hundred million years of conserved synteny between chicken Z and human chromosome 9, *Nature Genetics*, 21 (1999): 258–9.

W. J. Swanson and V. D. Vacquier, The rapid evolution of reproductive proteins, *Nature Reviews Genetics*, 3 (2002): 137–44.

C. A. Tilford *et al.*, A physical map of the human Y chromosome, *Nature*, 409 (2001): 943–5.

M. F. Wolfner, The gifts that keep on giving: physiological functions and evolutionary dynamics of male seminal proteins in *Drosophila*, *Heredity*, 88 (2002): 85–93.

G. J. Wyckoff, W. Wang and C.-I. Wu, Rapid evolution of male reproductive genes in the descent of man, *Nature*, 403 (2000): 304–8.

D. Zarkower, Establishing sexual dimorphism: conservation amidst diversity?, *Nature Reviews Genetics*, 2 (2001): 175–85.

Chapter 2: The Common Man

M. F. Fathalla, The girl child, *Int. J. Gyn. Obst.*, 70 (2000): 7–12.

D. Gil *et al.*, Male attractiveness and differential testosterone investment in zebra finch eggs, *Science*, 286 (1999): 126–8.

R. Jacobsen, H. Moller and A. Mouritsen, Natural variation in the human sex ratio, *Human Reprod.*, 14 (1999): 3120–5.

W. H. James, The association between the sexes of children within sibships, *Human Reprod.*, 15 (2000): 1422.

A. Malpani, A. Malpani and D. Modi, Preimplantation selection for family balancing in India, *Human Reprod.*, 17 (2002): 11–12.

R. G. Nager *et al.*, Experimental demonstration that offspring sex ratio varies with maternal condition, *Proc. Natl. Acad. Sci. USA*, 96 (1999): 570–3.

C. Packer, D. A. Collins and L. E. Eberly, Problems with primate sex ratios, *Phil. Trans. R. Soc. B*, 355 (2000): 1627–35.

R. Stouthamer, J. A. J. Breeuwer and G. D. D. Hurst, *Wolbachia pipentis*; microbial manipulator of arthropod reproduction, *Ann. Rev. Microbiol.*, 53 (1999): 71–102.

S. Sudha and S. I. Rajan, Female demographic disadvantage in India 1981–1991: sex selective abortions and female infanticide, *Development and Change*, 30 (1999): 585–618.

J. H. Werren and L. W. Beukeboom, Sex determination, sex ratios, and genetic conflict, *Ann. Rev. Ecol. Syst.*, 29 (1998): 233–61.

S. A. West, S. E. Reece and B. C. Sheldon, Sex ratios, *Heredity*, 88 (2002): 117–24.

Chapter 3: Seven Ages of Mankind

M. Beato and J. Klug, Steroid hormone receptors: an update, *Human Repro. Update*, 6 (2000): 225–36.

R. G. Bribiescas, Reproductive ecology and life history of the human male, *Ybk. Phys. Anthropol.*, 44 (2001): 148–76.

M. H. Choi, Y. S. Yoo and B. C. Chung, Biochemical roles of testosterone and epitestosterone to 5-alpha reductase as indicators of male-pattern baldness, *J. Invest. Dermatol.*, 116 (2001): 57–61.

S. R. Davis and J. Tran, Testosterone influences libido and well-being in women, *Trends Endocrinol. Metab.*, 12 (2001): 33–7.

A. D. Dreger, *Hermaphrodites and the Medical Invention of Sex* (Cambridge, Mass.: Harvard Univ. Press, 1998).

J. A. Ellis, M. Stebbing and S. B. Harrap, Polymorphism of the androgen receptor gene is associated with male pattern baldness, *J. Invest. Dermatol.*, 116 (2001): 452–5.

M. Foucault, *Herculine Barbin: Being the Recently Discovered Memoirs of a Nineteenth Century French Hermaphrodite* (New York: Pantheon, 1980).

D. Gould, R. Petty and H. S. Jacobs, The male menopause: does it exist?, *Brit. Med. J.*, 320 (2000): 858–61.

F. Kandeel *et al.*, Male sexual function and its disorders: physiology, pathophysiology, clinical investigation, and treatment, *Endoc. Rev.*, 22 (2001): 342–88.

S. A. Kidd *et al.*, Effects of male age on semen quality and fertility: a review of the literature, *Fert. Steril.*, 75 (2001): 237–49.

J. T. Martin, Sexual dimorphism in immune function: the role of prenatal exposure to androgens and estrogens, *Euro. J. Pharmacol.*, 405 (2000): 251–61.

H. F. L. Meyer-Bahlburg *et al.*, Gender change from female to male in classical congenital adrenal hyperplasia, *Hormones Behav.*, 30 (1996): 319–22.

I. E. Messinis and S. D. Milingos, Leptin in human reproduction, *Human Repro. Update*, 5 (1999): 52–63.

S. Nef and L. F. Parada, Hormones in male sexual development, *Genes Dev.*, 14 (2000): 3075–86.

K. L. Parker, A. Schedl and B. P. Schimmer, Gene interactions in gonadal development, *Ann. Rev. Physiol.*, 61 (1999): 417–33.

L. Pinsky, R. P. Erickson and R. N. Schimke, *Genetic Disorders of Human Sexual Development* (Oxford: Oxford Univ. Press, 1999).

J. E. Schneider, D. Zhou and R. M. Blum, Leptin and metabolic control of reproduction, *Horm. Behav.*, 37 (2000): 306–26.

K. Segrave, *Baldness: A Social History* (Jefferson, NC: McFarland, 1996).

P. J. Wang *et al.*, An abundance of X-linked genes expressed in spermatogonia, *Nature Genetics*, 27 (2001): 422–6.

Chapter 4: Hydraulics for Boys

L. A. Aytac *et al.*, Socioeconomic factors and incidence of erectile dysfunction: findings of the longitudinal Massachusetts Male Aging Study, *Soc. Sci. Med.*, 51 (2000): 771–8.

P. Bennet and V. A. Rosairio, *Solitary Pleasures: The Historical, Literary and Artistic Discourses of Autoeroticism* (London: Routledge, 1995).

D. G. Hatzichristiou and E. S. Pescatori, Current treatment and emerging therapeutic approaches in male erectile dysfunction, *Brit. J. Urol. Internat.*, 88/suppl. 3 (2001): 11–17.

W. Insmore and C. Evans, ABC of sexual health: erectile dysfunction, *Brit. Med. J.*, 318 (1999): 387–90.

R. Kirby, C. Carson and I. Goldstein, *Erectile Dysfunction: A Clinical Guide* (Oxford: Isis Medical Media, 1999).

G. Wagner and J. Mulhall, Pathophysiology and diagnosis of male erectile dysfunction, *Brit. J. Urol. Internat.*, 88/suppl. 3 (2001): 3–10.

Chapter 5: Man Mutilated

Z. Farshi, K. R. Atkinson and R. Squire, A study of clinical opinion and practice regarding circumcision, *Arch. Dis. Childhood*, 83 (2000): 393–6.

A. R. Favazza, *Bodies Under Siege: Self-Mutilation in Culture and Psychiatry* (Baltimore: Johns Hopkins Univ. Press, 1996).

R. Goldman, The psychological impact of circumcision, *Brit. J. Urol. Internat.*, 83/suppl. 1 (1999): 93–102.

D. Gollaher, *Circumcision: A History of the World's Most Controversial Surgery* (New York: Basic Books, 2000).

D. T. Halperin and R. C. Bailey, Male circumcision and HIV infection: 10 years and counting, *Lancet*, 354 (1999): 1813–15.

T. Hammond, A preliminary poll of men circumcised in infancy or childhood, *Brit. J. Urol. Internat.*, 83/suppl. 1 (1999): 85–92.

K. O'Hara and J. O'Hara, The effect of male circumcision on the sexual enjoyment of the female partner, *Brit. J. Urol. Internat.*, 83/suppl. 1 (1999): 79–84.

A. Rosler and E. Witztum, Treatment of men with paraphilia with a long-acting analogue of gonadotropin-releasing hormone, *New Engl. J. Med.*, 338 (1998): 416–22.

E. J. Schoen, Benefits of newborn circumcision: is Europe ignoring medical evidence?, *Arch. Dis. Childhood*, 77 (1997): 258–60.

P. G. Schwingl and H. A. Guess, Safety and effectiveness of vasectomy, *Fert. Steril.*, 73 (2000): 923–36.

G. Taylor, *Castration: An Abbreviated History of Western Manhood* (London: Routledge, 2000).

Chapter 6: Bois-Regard's Worms

J. K. Amory and W. J. Bremner, Newer agents for hormonal contraception in the male, *Trends Endocrinol. Metab.*, 11 (2000): 61–6.

A. W. S. Cha *et al.*, Foreign DNA transmission by ICSI: injection of spermatozoa bound with exogenous DNA results in embryonic GFP expression and live Rhesus monkey births, *Molec. Hum. Repro.*, 6 (2000): 26–33.

L. V. Depaolo, B. T. Hinton and R. E. Braun, Male contraception: views to the 21st century, *Trends Endocrinol. Metab.*, 11 (2000): 66–9.

J. P. Evans, Getting sperm and egg together: things conserved and things diverged, *Biol. Repro.*, 63 (2000): 355–60.

M. J. G. Gage, Mammalian sperm morphometry, *Proc. Roy. Soc. Lond. B*, 265 (1998): 97–103.

M. Kayser *et al.*, Characteristics and frequency of germline mutations at microsatellite loci from the human Y chromosome, as revealed by direct observation in father/son pairs, *Am. J. Hum. Genet.*, 66 (2000): 1580–8.

R. C. Kuo *et al.*, NO is necessary and sufficient for egg activation at fertilization, *Nature*, 406 (2000): 633–6.

D. J. Mclean *et al.*, Germ cell transplantation and the study of testicular function, *Trends Endoc. Metab.*, 12 (2001): 16–21.

T. Ogawa *et al.*, Transplantation of male germ line stem cells restores fertility in infertile mice, *Nature Medicine*, 6 (2000): 29–34.

B. T. Preston *et al.*, Dominant rams lose out by sperm depletion, *Nature*, 409 (2001): 681–2.

T. Tregenza, Evolutionarily dynamic sperm, *Trends Res. Evol. Ecol.*, 15 (2000): 85–6.

P. Wassarman, L. Jovine and E. S. Litshcer, A profile of fertilization in mammals, *Nature Cell Biol.*, 3 (2001): 59–64.

Chapter 7: Bend Sinister

M. Bamshad *et al.*, Genetic evidence on the origins of Indian caste populations, *Genome Research*, 11 (2001): 994–1004.

D. D. Brewer *et al.*, Prostitution and the sex discrepancy in the reported number of sexual partners, *Proc. Natl. Acad. Sci. USA*, 97 (2000): 12385–8.

D. R. Carvalho-Silva *et al.*, The phylogeography of Brazilian Y-chromosome lineages, *Amer. J. Hum. Genet.*, 68 (2000): 281–6.

D. Einon, How many children can one man have?, *Evol. Hum. Behav.*, 19 (1998): 413–26.

M. A. Jobling, In the name of the father: surnames and genetics, *Trends Genet.*, 17 (2001): 353–7.

A. M. Johnson *et al.,* Sexual behaviour in Britain: partnerships, practices and HIV risk behaviours, *Lancet,* 358 (2001): 1835–42.

A. McWhinnie, Gamete donation and anonymity: should offspring from donated gametes continue to be denied knowledge of their origins and antecedents?, *Human Reprod.*, 16 (2000): 807–17.

G. Reekie, *Measuring Immorality: Social Enquiry and the Problem of Illegitimacy* (Cambridge: Cambridge Univ. Press, 1998).

B. Sykes and C. Irven, Surnames and the Y chromosome, *Amer. J. Hum. Genet.*, 66 (2000): 1417–19.

Chapter 8: James James's Skull

G. Barbujani and G. Bertorelle, Genetics and the population history of Europe, *Proc. Natl. Acad. Sci. USA*, 98 (2001): 22–5.

J. Bertranpetit, Genome, diversity, and origins: the Y chromosome as a storyteller, *Proc. Natl. Acad. Sci. USA*, 97 (2000): 6927–9.

R. L. Cann, Genetic clues to dispersal in human populations: retracing the past from the present, *Science,* 291 (2001): 1742–8.

M. Gullestad and M. Segalne, *Family and Kinship in Europe* (London: Pinter, 1997).

E. W. Hill, M. A. Jobling and D. G. Bradley, Y-chromosome variation and Irish origins, *Nature*, 404 (2000): 351.

L. Jorde *et al.*, The distribution of human genetic diversity: a comparison of mitochondrial, autosomal and Y-chromosome data, *Amer. J. Hum. Genet.*, 66 (2000): 979–88.

T. M. Karafet *et al.*, Ancestral Asian source(s) of New World Y-chromosome founder haplotypes, *Amer. J. Hum. Genet.*, 64 (1999): 817–31.

J. T. Lell and D. C. Wallace, The peopling of Europe from the maternal and paternal perspectives, *Amer. J. Hum. Genet.*, 67 (2000): 1376–81.

F. R. Santos *et al.*, The central Siberian origin for Native American Y chromosomes, *Amer. J. Hum. Genet.*, 64 (1999): 619–28.

M. Seielstad, E. Minch and L. L. Cavalli-Sforza, Genetic evidence for a higher female migration rate in humans, *Nature Genetics*, 20 (1998): 278–80.

O. Semino *et al.*, The genetic legacy of Paleolithic *Homo sapiens sapiens* in extant Europeans: a Y chromosome perspective, *Science*, 290 (2000): 1155–9.

M. P. H. Stumpf and D. B. Goldstein, Genealogical and evolutionary inference with the human Y chromosome, *Science*, 291 (2001): 1738–42.

P. Underhill *et al.*, Y chromosome sequence variation and the history of human populations, *Nature Genetics*, 26 (2000): 358–61.

J. F. Wilson *et al.*, Genetic evidence for different male and female roles during cultural transitions in the British Isles, *Proc. Natl. Acad. Sci. USA*, 98 (2001): 5078–83.

Chapter 9: Polymorphous Perversity

T. Birkhead, *Promiscuity: An Evolutionary History of Sperm Competition and Conflict* (London: Faber, 2000).

A. Cockburn, Evolution of helping behavior in cooperatively breeding birds, *Ann. Rev. Ecol. Syst.*, 29 (1998): 141–77.

A. F. Dixson, *Primate Sexuality: Comparative Studies of the Prosimians, Monkeys, Apes and Human Beings* (Oxford: Oxford Univ. Press, 1998).

M. Eens and R. Pinxten, Sex role reversal in vertebrates: behavioural and endocrinological accounts, *Behav. Processes*, 51 (2000): 135–47.

M. Emond and J. Scheib, Why not donate sperm? A study of potential donors, *Evol. Hum. Behav*, 19 (1998): 313–19.

M. D. Jennions and M. Petrie, Why do females mate multiply? A review of the genetic benefits, *Biol. Rev. Camb. Phil. Soc*, 75 (2000): 21–64.

T. L. Karr and S. Pitnick, Sperm competition: defining the rules of engagement, *Curr. Biol.*, 9 (1999):787–90.

S. L. Klein, The effects of hormones on sex differences in infection: from genes to behavior, *Neurosci. Behav. Rev.*, 24 (2000): 627–38.

J. C. Mitani, D. P. Watts and M. N. Muller, Recent developments in the study of wild chimpanzee behaviour, *Evol. Anthropol.*, 11 (2002): 9–25.

J. M. Plavcan, Sexual dimorphism in primate evolution, *Ybk. Phys. Anthropol.*, 44 (2001): 25–53.

A. C. Stone *et al.*, High levels of Y-chromosome nucleotide diversity in the genus *Pan*, *Proc. Natl. Acad. Sci. USA*, 99 (2002): 43–8.

R. Wrangham and D. Peterson, *Demonic Males: Apes and the Origins of Human Violence* (New York: Houghton Mifflin, 1996).

Chapter 10: A Martian on Venus

E. Aries, *Men and Women in Interaction: Reconsidering the Differences* (Oxford: Oxford Univ. Press, 1996).

E. Badinter, *XY: De l'identité masculine* (Paris: Odile Jacob, 1992).

S. A. Berenbaum, Effects of early androgens on sex-typed activities and interests in adolescents with congenital adrenal hyperplasia, *Hormones Behav.*, 35 (1999): 102–10.

T. Birkhead, H. Schwabl and T. Burke, Testosterone and maternal effects: integrating mechanisms and functions, *Trends Res. Evol. Ecol.*, 15 (2000): 86–7.

J. M. Dabbs and M. G. Dabbs, *Heroes, Rogues and Lovers: Testosterone and Behavior* (New York: McGraw-Hill, 2000).

A. Fausto-Sterling, *Sexing the Body: Gender Politics and the Construction of Sexuality* (New York: Basic Books, 2000).

J. N. Giedd *et al.*, Brain development during childhood and adolescence: a longitudinal MRI study, *Nature Neuroscience*, 2 (1999): 861–3.

S. Golombok and R. Fivush, *Gender Development* (Cambridge, Cambridge Univ. Press, 1994).

B. Low, *Why Sex Matters: A Darwinian Human Behavior* (Princeton: Princeton Univ. Press, 1999).

M. J. Maynes *et al.*, *Gender, Kinship, Power: A Comparative Interdisciplinary History* (New York: Routledge, 1996).

L. Mealey, *Sex Differences: Development and Evolutionary Strategies* (San Diego: Academic Press, 2000).

C. L. Ridgway and L. Smith-Lovin, The gender system and interaction, *Ann. Rev. Sociol.*, 25 (1999): 191–216.

L. Stone, *Kinship and Gender: An Introduction* (Boulder: Westview, 1997).

Envoi: The Descent of Men

L. Bujan *et al.*, Increase in scrotal temperature in car drivers, *Human Reprod.*, 15 (2000): 1355–7.

A. Cagnacci, N. Maxia and A. Volpe, Diurnal variation of semen quality in human males, *Human Reprod.*, 14 (1999): 106–9.

A. Clare, *On Men: Masculinity in Crisis* (London: Arrow Books, 2001).

E. M. Crimmins and Y. Saito, Trends in healthy life expectancy in the United States, 1970–1990: gender, racial and educational differences, *Soc. Sci. Med.*, 52 (2001): 1629–41.

A. Hughes, How vulnerable is the developing testis to the external environment?, *Arch. Dis. Childhood*, 83 (2000): 281–2.

D. S. Irvine, Male reproductive health: cause for concern?, *Andrologia*, 32 (2000): 195–208.

N. Jorgensen *et al.*, Regional differences in semen quality in Europe, *Human Reprod.*, 16 (2001): 1012–19.

S. Kraemer, The fragile male, *Brit. Med. J.*, 321 (2000): 1609–12.

P. McDonough and V. Walters, Gender and health: reassessing patterns and explanations, *Soc. Sci. Med.*, 52 (2001): 547–59.

T. O'Dowd and D. Jewell, *Men's Health* (Oxford: Oxford Univ. Press, 1998).

C.-J. Partsch, M. Aukamd and W. G. Sippell, Scrotal temperature is increased in disposable plastic lined nappies, *Arch. Dis. Childhood*, 83 (2000): 364–8.

R. F. A. Weber *et al.*, Environmental influences on male reproduction. *Brit. J. Urol. Internat.*, 89 (2002): 143–8.

INDEX